AGLOW IN THE DARK

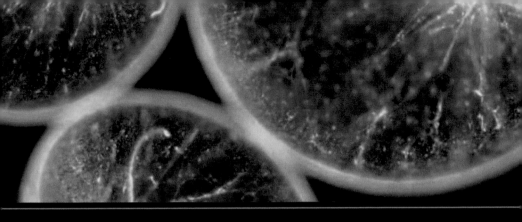

VINCENT PIERIBONE AND DAVID F. GRUBER

AGLOW

THE BELKNAP PRESS OF HARVARD UNIVERSITY PRESS

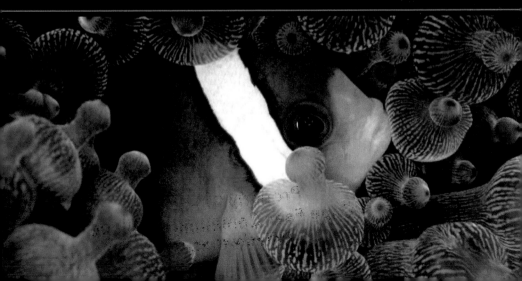

The Revolutionary Science of Biofluorescence

IN THE DARK

CAMBRIDGE, MASSACHUSETTS, AND LONDON, ENGLAND   2005

Chapter-opening illustration: *Aequoria victoria*, © Amy Bartlett Wright, 2005.
Repeating ornament in text: *Aurelia aurita*, from web page *www.glf.dfo-mpo.gc.ca*,
courtesy of Fisheries and Oceans Canada. Reproduced with the permission of
Her Majesty the Queen in Right of Canada, 2005.
Design: Gwen Nefsky Frankfeldt

Library of Congress Cataloging-in-Publication Data

Pieribone, Vincent.
Aglow in the dark : the revolutionary science of biofluorescence /
Vincent Pieribone and David F. Gruber.
    p.   cm.
Includes bibliographical references and index.
ISBN 0-674-01921-0 (hardcover : alk. paper)
1. Bioluminescence.   I. Gruber, David F.   II. Title.

QH641.P54   2005
572'.4358—dc22   2005048050

CONTENTS

FOREWORD

· · · · · · · · · · · · · · · · · · · ·

Walking in space for the first time and surrounded by "luminous objects that glowed in the black sky," the astronaut John Glenn recalled the friendly twinkling of fireflies from summer nights in Ohio when he was growing up. Walking through the woods during the Stone Age, natives of the West Indies tucked shimmering beetles between their toes to make the dark seem less scary. Throughout history, human beings have been captivated by the cool glow of other living creatures. But although an extraordinary number of animals, from sharks to microscopic dinoflagellates, are capable of lighting up, how they do it remained shrouded in mystery long after Newton published *Opticks*. It was not until the turn of the last century that a French physiology professor deciphered the basic chemistry of bioluminescence. Yet a mere three or four scientific generations later, biomedical researchers are using genetically modified fluorescent proteins to light up the interior of *living* cells.

The implications of using "light by nature" to illuminate the secrets of life are simply staggering. A rainbow of fluorescent proteins is throwing open to exploration a living terra incognita that is beyond the reach of even the most powerful microscopes and scanners. It now seems entirely plausible that this new territory—the deepest and hitherto darkest chambers of cells—will prove to be every bit as vast and full of promise (and dangers too, no doubt) as those opened up in the nineteenth century by the locomotive, steamship, and telegraph. Fluorescent proteins that can be injected into virtually any living cell are now essential tools for furthering the molecular revolution and fighting diseases like AIDS, Alzheimer's, and cancer. Like other world-changing inventions, this tool has other as yet untapped uses. Apparently new versions of these proteins may soon allow people to telegraph their thoughts to computers as if they were directing their own hands or eyes.

Aglow in the Dark explains all this and more. Written by two scientists in sparkling prose, the book is an authoritative introduction to the science of fluorescent proteins, a concise natural history of bioluminescence, and, last but hardly least, a highly readable and hugely absorbing discovery saga full of quirky characters, ironic twists, and chance encounters—a *Microbe Hunters* for modern sensibilities.

I was most fascinated, I have to admit, by the authors' wonderfully intimate, completely authentic portrait of a unique creative community. The chance to learn how inventive people in an entirely different discipline play their "games against nature" and pursue their versions of the mathematician John Nash's "truly original idea" was incredibly stimulating and instructive. Of course, the rules, strategies, and environments of biologists differ dramatically from those of biographers. But *Aglow in the Dark* highlights what they have in common: As the lucky inhabitants of this Information Age, we are *all* trying to generate our *own* light—and isn't that precisely what ideas are?

Pieribone and Gruber weave myriad dates, facts, explanations, anecdotes, observations, and bon mots into a seamless coming-of-age narrative, tracing how the science of bioluminescence evolves from pure speculation to description and, in the 1990s, invention. It made me wonder, of course, about the environmental stimuli. In the nonhuman species that evolved the ability to glimmer and glow, luminescence can be, among other things, a defensive reaction, a sexual come-on, or a lure with which to attract lunch. When it comes to nature's light shows, Pieribone and Gruber report no instances of art for art's sake. Romance and wanderlust motivated individual scientists like Princeton's Edmund Harvey. But, for the most part, creative thinking about luminescence seemed to wax and wane over the past century with possibilities of adapting it for a variety of practical human uses.

The first major breakthrough apparently coincided with the rise of Darwinism, commercial culture, and the chemical industry, those catalysts for a wholesale ransacking of nature for marketable ideas. Raoul Dubois's ingenious display of sconces filled with *Vibrio* bacteria at the Paris exhibition of 1900 was clearly part of a Klondike-style stampede to find cheaper, safer, brighter ways to light homes and workplaces. The next strides were made by a heroically stubborn survivor of Nagasaki during the Cold War years when the military-industrial complex set the scientific agenda. During World War II, Japanese military planners had briefly toyed with a primitive precursor of night vision goggles, but the American navy, like their Soviet counterpart, was more interested in the problems posed by swarms of plankton that float on the waves and glitter when disturbed. That glitter can disorient navigators, camouflage enemies, or turn a submarine lurking beneath the surface into a highly visible target.

The explosion that turned bioluminescence from a descriptive discipline into a rich fount of invention didn't really occur until the 1990s when the information technology and genome revolutions sparked a search for fluorescent proteins that could be reengineered and mass-produced. Supply and demand took off in tandem. Cloning made it possible to copy specimens without wasting time gathering and drying thousands of, say, jellyfish. The more colors and the brighter the light, the more uses researchers found for them. The faster the applications grew, the greater the clamor for new, improved proteins. Proteins with red fluorescence, brighter fluorescence, and sensitivity to electrical charges were just some of the by-products.

So what attracted these pioneers? The game, apparently. In *The Mathematician's Apology*, G. H. Hardy famously claimed that mathematics, indeed all science, is a competitive sport, and no one who has read *The Double Helix*, James Watson's account of his and Francis Crick's race to beat Linus Pauling to DNA, will probably ever think otherwise. Tournaments are marvelously efficient arrangements from society's point of view—the ultimate cheap date, Joseph Schumpeter, the Viennese polymath and author of one of the first general theories of creativity, notes. They elicit a maximum of effort from individuals at minimal societal cost. For the sake of a few really big prizes—a Nobel Prize, a rich patent, or an endowed chair—many talented people will enter the race and make great personal sacrifices. But, even among those who stick around for the finish, only a handful can collect gold. *Aglow in the Dark* is full of researchers who join the race, make critical contributions, and then disappear from the historical narrative of bioluminescence. The only woman described in the book, a Columbia graduate student, is one such scientist. The Woods Hole researcher who deciphered the sequence of green fluorescent protein is another. He moved on to study mosquitoes. Asked why he bothered with green fluorescent protein in the first place, he replied, "Because I'm weird." At one point, Roger Tsien, the entrepreneur in the story, asks a Berkeley colleague, "Do you realize what you have here? . . . A gold mine!" He quickly realized that his colleague had other aims. "I got the impression that this greed of mine had not occurred to him."

The potential for making money, the authors make clear, doesn't mean that money is the main motivation even for a go-getter like Tsien. In science, as in business or the movies, cash is a convenient, if imperfect, way to keep score—an enabler, a way to reimburse the family for the nights and weekends spent away from them. The currency that the scientists in these pages, like scholars everywhere, crave most is peer recognition. Although Tsien holds dozens of patents, he claims that his work is "often dismissed as technology development, inferior to pure biology." What he doesn't say is that "pure biology" remains the gold standard for judging achievement among the Swedish scientists who sit on the Nobel Prize committees.

If producing ideas is a competitive sport, what do the winners have in common? Some of the same things as creative thinkers in other fields, it turns out: Getting into the game early. Playing for high stakes. Total concentration. Sticking to a game plan. There is one element of surprise: Despite a few fortuitous spills in the lab sink, strategy—such as choosing an approach that nobody else has tried—rather than serendipity seems to determine the outcome more often than not. One researcher's remark that he was happy to be working in an unpopular field because "you don't find anything new by trying what everyone else already has" reminded me of John Nash's research strategy: While everyone else who wanted to get to a peak looked for a path somewhere on the mountain, Nash would scale a different peak in order to shine a searchlight back on the first one. There is also an apparent paradox: Intense competition seems to breed an extraordinary amount of cooperation. The researchers are constantly exchanging ideas and forming new coalitions. Or maybe it's not paradoxical at all. Technology companies also share patents with rivals, form research consortia, and so on.

Happily, *Aglow in the Dark* shows that races are still games. Even though bioluminescence has grown up, silliness, playfulness, and other vestiges of adolescence remain. Tsien's brilliant career seems to have started, like John Nash's, with a bang—an attempt to mix explosives in the family basement. The head of a major lab made his scientific debut atop a table reciting a bowdlerized version of *King Lear* while clad in bathing trunks and a tie. The Russian maverick who beat everyone else to the holy grail of red fluorescence is addicted to video games and has his whole eccentric extended family on the payroll. In the course of their far-flung reporting, Pieribone and Gruber went diving in the Great Barrier Reef, watched Pixar's hit movie, *Finding Nemo*, in New York City, and took a guided tour of neon aquariums in apartments all over Moscow. The authors convey the spirit of the enterprise perfectly. They leave the reader with the sense that, everything being connected, life is no less likely to yield its secrets to someone who spends a lifetime studying a glowworm than to one who travels to the stars. It's not only a race, the authors seem to be saying, but also a vision.

Sylvia Nasar
Yaddo

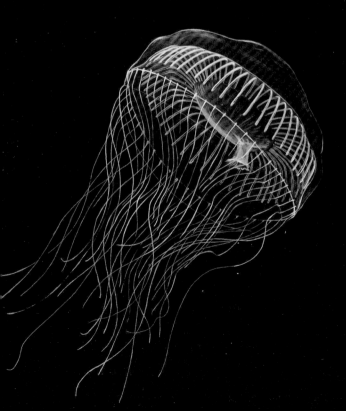

Prologue

I N NOVEMBER 2004, inside a darkened room on the fifth floor of the Skirball Institute at New York University Medical School, a neurobiologist begins an experiment on the inner workings of the brain. The researcher sits in front of a two-photon microscope while a three-month-old mouse, its heart beating calmly under anesthesia, rests its head below the gaze of the machine. A small circular incision in its scalp reveals a milky white skull. The microscope lens is gently lowered until a blurry image of the blood vessels covering the brain surface appears. The scientist focuses on a patch of featureless pinkish tissue within the living brain. This is the cerebral cortex, the outer layer of the brain responsible for higher functions such as comprehension, memory, and complex movements. It is the most elaborate and sophisticated tissue in nature. One square millimeter of brain tissue contains thousands of individual nerve cells in an intricately woven lattice. Like a living three-dimensional computer chip, every nerve cell acts as a tiny electronic processing machine.

Working by the glow of the video monitor, the scientist clicks the scan button and suddenly the microscope springs to life with the sounds of shutters clicking and motors purring. Unlike conventional microscopes, this machine is equipped with a powerful infrared laser, initially developed by the defense industry to intercept nuclear warheads in space. It emits a high-energy, high-density beam of light that delicately scans the brain tissue without damaging it, by firing at extremely short pulses, each flash lasting a thousand times less than a trillionth of a second. Within moments, glowing green figures materialize on the screen, forming a crisp and detailed picture of the mouse's neurons. If this were a normal mouse, the screen would be blank, but this mouse is far from ordinary. It has been genetically engineered to produce a unique fluorescing protein found in a vanishing species of jellyfish. The protein fills the nerve cells, causing them to glow brightly when excited by the laser beam. Elegant glowing treelike structures emerge as the lens moves, giving a three-dimensional tour of the mouse's brain. Images appear of nerve cells bearing intricate vinelike appendages. These, in turn, sprout foliate protrusions, synapses, which are the foundation of brain function. For a mouse, they store information such as a favorite smell, the location of a nest, or the sound of approaching danger.

As the researcher ventures deeper into this glowing Lilliputian forest, a lifeless zone appears: a tangled briar patch of twisted and dying branches buried within the healthy brain. At its core resides a dense mass of dead material. This is an Alzheimer's plaque, a grotesque scar in otherwise vibrant brain tissue. It is the pathological hallmark of one of the most prevalent brain disorders afflicting humans. As Alzheimer's disease progresses, these plaques litter the gray matter of the brain, slowly ravaging it. Alzheimer's disease is characterized by a progressive and irreversible

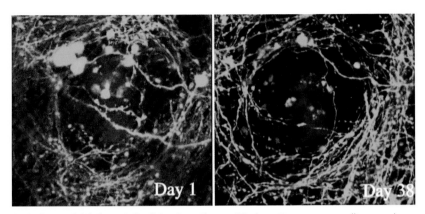

A single amyloid deposit (red) is viewed over 38 days. Some nerve cell appendages (green) near the deposit are killed. Photo by Julia Tsai and Wen Biao Gan.

memory loss that eventually results in dementia. In early stages of the disease, the patient begins having difficulty completing simple tasks, such as balancing a checkbook, driving a car, or operating household appliances. Speech becomes more difficult as words are harder to find. At the onset, the disease is often overlooked, the symptoms being dismissed as normal signs of aging.

But as Alzheimer's disease progresses, patients increasingly neglect daily habits, such as bathing and grooming. They have difficulty remembering new information, and memories start becoming warped and distorted. Language skills deteriorate, and patients utter incomplete and incomprehensible sentences. They start exhibiting restless behavior such as mood swings, anxiety, confusion, and paranoia, as well committing aggressive physical or verbal acts. Memories soon crumble to the point that the afflicted, heartbreakingly, fail to recognize even beloved spouses and children.

At the terminal stage of the disease, the patient becomes incontinent and dependent on caregivers. All semblance of the patient's personality, stripped of identity and self, vanishes.

Alzheimer's disease is not naturally present in mice, but this mouse has been genetically engineered to express the human mutated form of amyloid precursor protein and presenilin I—the building blocks of toxic plaques. Together, these mutations produce an animal model of Alzheimer's disease. By three months of age, the mouse will develop an analogue of the human condition, losing its ability to store memories and impairing its ability to run mazes. The researcher begins by examining a healthy and pristine mouse brain, but over time she watches the filigree of neurons become emaciated and lose their contact with neighboring neurons. Some neurons are completely engulfed by burgeoning plaques. Over the next weeks, her work will reveal that these plaques grow and spread, providing a direct account, for the first time, of how this debilitating disease progressively destroys a healthy brain. The same technique can also be used to screen for drugs capable of reversing or halting the degradation caused by Alzheimer's disease.

Once considered science fiction, journeys into the living brain are becoming commonplace. The illumination power afforded by green fluorescent protein (GFP), in conjunction with genetic engineering, is dramatically transforming most areas of modern biomedical science. The same protein that is being used to tackle Alzheimer's disease may one day enable scientists to directly link the human brain to computers, providing the ability to move objects and even fly fighter jets merely by thinking. The applications of such technologies, driven by this simple glowing protein, are extending the boundaries of science, allowing researchers to understand, manipulate, and interact with the living brain. The scientific possibilities, uniquely beneficial and potentially nefarious, are seemingly endless.

CHAPTER ONE

Living Light

BIOLUMINESCENCE, the ability of living creatures to generate light, may seem a novelty, a freak of nature, but lurking in the darkest portions of the planet, it is commonplace. If we divide all life on Earth into phyla, groupings based on their morphology and physical characteristics, almost half have representatives that glow.[1] This diverse set of creatures includes bacteria, protozoa, fungus, molds, jellyfish, insects, squid, worms, crustaceans, mollusks, and sharks. Bioluminescence, or living light, is a phenomenon that has captivated people for thousands of years. One of the earliest written accounts dates back to Aristotle, almost 2,500 years ago. In *On Colours* he wrote, "Some things which are neither fire nor forms of fire seem to produce light by nature."[2] Here Aristotle correctly distinguished between light radiated from hot objects—known as incandescence—and light generated without heat, or luminescence. Practically any solid body that can be heated to 525 degrees Celsius will emit a faint dull red glow. As the temperature increases, the color changes to a cherry red, then yellow, and finally white. Bioluminescence is a chemical reaction that burns fuel and releases light with such perfect efficiency that it barely produces heat—a cold light. Conversely, a light bulb uses only a small percentage of its energy to produce light, the rest being released as heat. Working within the limitations of his understanding, Aristotle made many astute observations but faltered in his explanations of what he calls "other lights." In *De Sensu and De Memoria* he says: "It is the nature of smooth things to shine in the dark," and he gives as examples the "heads of certain fishes and the juice of the cuttle-fish."[3] We now know that specialized glands in cuttlefish produce bioluminescence and that the shining fish heads owe their glow to luminescent bacteria growing on their decaying flesh.

In the first century Gaius Plinius Secundus, a Roman statesman, naturalist, and writer known as Pliny the Elder began to document some of the glowing creatures near his home on the slopes of Mount Vesuvius (before perishing in its eruption in the year 79). There, he had access to the Bay of Naples, a body of water home to many bioluminescent organisms. He identified several glowing animals, including one that he called "Plumo Marinus," a large jellyfish recognized today as *Pelagia noctiluca*, or the purple jellyfish. He also remarked on the edible luminescent clam, *Pholas dactylus.* This marine bivalve, found primarily in Europe, drills itself into soft rock and squirts a blue luminous liquid from its siphon when disturbed. At night, ejaculating clams uncovered by the tide provide a twinkling light display. Pliny noted that the clams continued to shine even as they were eaten. "It is the nature of these fish to shine in darkness with a bright light when other light is removed, and in proportion to their amount of moisture to glitter both in the mouth of persons masticating them and in their hands, and even on the floor and on their clothes when drops fall from them, making it clear beyond all doubt that their juice possesses a property that we should marvel."[4] Pliny also relates the first recorded human use of bioluminescence, describing how he used a walking stick rubbed against a jellyfish's slime to light up his trail like a torch.

During the next 1,300 years, an era of scant original scientific observation, mentions of bioluminescent organisms in literature are scarce, probably because such eerie creatures were often considered taboo. Italians had a strong superstitious dread of fireflies, believing them to be the spirits of their departed ancestors.[5] Dante was one of the few Italians fearless enough to mention bioluminescence in his prose. In the *Inferno*, one of the last poems he penned before his death in 1321, he wrote of peering down into the Eighth Chasm of Hell and seeing "fireflies innumerous spangling o'er the vale."[6]

It wasn't until the seventeenth century that a more scientific approach to the study of glowing creatures emerged. Robert Boyle, a founder of the Royal Society of London, was part of a new breed of researchers/philosophers who based their conclusions on what Boyle called "interrogations of nature." In late 1667, while living in Oxford, Boyle used a crude air pump to briefly remove air from a bell jar containing a bioluminescent fungus. He found that without air the bioluminescence disappeared. When air returned, the fungus glowed once again. Thus Boyle discovered the first chemical feature of bioluminescence: it requires air. Boyle reported his results in a December 16, 1672 paper to the *Transactions of the Royal Society*. At that time, the composition of "air" was still unknown. Later researchers would discover it is oxygen—which constitutes about a fifth of our atmosphere—that is required for bioluminescence.

Following Boyle's observation on the requirement for air, little progress on understanding the nature of living light was made during the eighteenth and first half of the nineteenth centuries. A prominent scientist, speaking before the Royal Society of London in 1810, summed up the situation:

> Many writers have ascribed the light of the sea to other causes than luminous animals. Martine supposed it to be occasioned by putrefac-

tion, Silberschlag believed it to be phosphoric, Professor J. Mayer conjectured that the surface of the sea imbibed light, which it afterwards discharged. Bajon and Gentil thought the light of the sea was electric, because it was excited by friction . . . I shall not trespass on the time of the Society to refute the above speculations; their authors have left them unsupported by either arguments or experiments, and they are inconsistent with all ascertained facts upon the subject.[7]

Bioluminescence research reached a turning point in 1887 when Raphaël Dubois, professor of physiology at the University of Lyons, France, and director of the Marine Laboratory at Tamaris-sur-Mer, discovered that two chemical components are necessary for bioluminescence. Dubois wrote that these "substances are necessary and sufficient to produce in vitro the phenomenon of luminescence, but the mechanisms had not been explained by any good hypothesis up until now."[8] A very thorough man, Dubois published a 275-page magnum opus on the luminescent properties of *Pyrophorus,* an elaterid beetle common in the West Indies.[9] He was initially attracted to this insect when he observed that even newly hatched larvae had petite luminescent organs, and he wanted to understand how the luminescence persisted long after the beetle perished. Dubois was aware of the numerous creative uses of the *Pyrophorus* bioluminescence. Natives of the West Indies, for example, sometimes stuck *Pyrophorus* beetles to their toes in order to illuminate a path through the woods at night, or used them to illuminate their homes, a practice that may have predated Pliny's description of bioluminescence use.

Dubois found that if he ground up the glowing parts of a dead beetle in cold water, the mixture would glow momentarily but then fade. If he ground up a beetle in boiling water, the mixture wouldn't glow at all, even upon cooling. To his amazement, Dubois found that if he added

Bioluminescent fungi of the genus *Panellus* on a tree branch. Photo by Osamu Shimomura.

the hot extracted mixture to the exhausted cold mixture (after its luminescence had faded), the mixture temporarily reignited. Any time he wanted the cold extract to glow again, he just needed to add additional hot extract. Dubois extended this finding to other species, starting with the luminescent edible clam, *Pholas dactylus*. From these simple experiments, Dubois drew two important conclusions. First, the light reaction requires two separate chemical components. Second, the fuel component of bioluminescence can withstand heating while the igniter, or catalyst, cannot. Dubois named these compounds after Lucifer, Latin for light-bearer. Following standard rules of nomenclature for biological

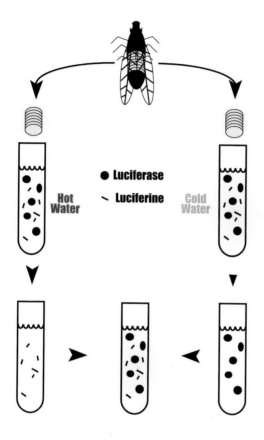

The famous Dubois experiment demonstrating the existence of a luciferase and luciferin.

compounds, he termed the catalyst "luciferase" and the fuel component "luciférine." The luciferin and luciferase principle appeared to hold true for many different glowing animals. Excited about his findings, Dubois designed a display for the Paris International Exposition of 1900 that used six 1-gallon flasks of bioluminescent bacteria to illuminate a room. Attendees entering the room found enough light to read a newspaper.

Dubois proposed using bioluminescence as a safe, cool way to light areas where explosives and other volatile substances are stored. He also proposed and (possibly patented) a miner's safety lamp made of luminescent bacteria. But these ideas never saw the light of day.

Bioluminescence is a relatively rare trait among land dwellers, but in the depths of the ocean more than 90 percent of animal species are capable of generating light.[10] Sunlight diminishes by a factor of 10 for every 75 meters of descent, as photons of light from the sun ricochet off particles and are absorbed by water.[11] Since the majority of the ocean's volume lies below the reach of the sun's rays, enveloped in perpetual darkness, marine creatures that can produce their own personal flashlights possess an advantage. Bioluminescence can aid in the search for prey; animals can either attract food with bioluminescent lures or use the light to scan the darkness. Bioluminescence can also deter an enemy with a blinding or distracting flash. And bioluminescent creatures can also use species-specific light shows to attract mates.

In the ocean's depths animals have evolved to employ bioluminescence in strange and mysterious ways. The female anglerfish, for example, is a voracious predator that lives in darkness 225 to 12,000 feet below the ocean's surface. The fish attracts prey by dangling and twitching a glowing orb in front of its gaping mouth. This lure is actually a modified dorsal fin spine containing a dense packet of bioluminescent bacteria cultivated by the fish inside a soft tissue bulb. When scientists first captured anglerfish, they noticed that all of the specimens were females, each about the size of an orange, and that each had what appeared to be one or a few large parasites attached to her body. Upon closer examination, it was discovered that these almond-sized hangers-on were actually male anglerfish. Males, attracted to the females by their odor and distinctive light displays, attach to a female, permanently. Over time many of

A bioluminescent deep-sea euphausiid shrimp *(Meganyctiphanes norvegica)*. The animal produces bioluminescence on its underside that perfectly matches the color and intensity of light from above, masking its shadow from predators. Photo by Edith Widder.

the males' organs, including their eyes and olfactory systems, degenerate and they begin to share blood vessels with the female. In some cases, several of the dwarf males attach to a single female. Their mouths become fused to the female's body, from which they derive nourishment.[12] In spite of their compromised position, the males continue to copulate with the female for the duration of their lives. Eventually every male degenerates into nothing more than a pair of testes that releases sperm

Left: A deep-sea anglerfish *(Melanocetus johnsoni)*, showing the bioluminescent lure (at the tip of the protruding stalk above the mouth) the fish uses to attract prey. Right: Close-up of a puny adult male anglerfish *(Linophryne indica)* fused to the side of a female. Photos copyright © 2005 Norbert Wu.

when the female releases hormones into her bloodstream which signal she is ready to release eggs. In his 1983 book, *Hen's Teeth and Horses Toes,* the Harvard biologist Stephen Jay Gould commented on this unusual co-habitation: "In some ultimate Freudian sense, what male could resist the fantasy of life as a penis with a heart, deeply and permanently embedded within a caring and providing female?"[13]

The most prominent bioluminescent organisms in the ocean are single-celled animals called dinoflagellates. (The name comes from the Greek for "whirling flagella," describing threadlike appendages the animal uses to propel itself through water.) The largest dinoflagellate, *Noctiluca scintillans*, a giant among the single-celled organisms, measures up to a millimeter in length, just visible to the unaided human eye. There are at least 1,800 species of dinoflagellates in the ocean, and as a group they are commonly regarded as planktonic algae. Yet some members are part-animal and part-plant. They possess chloroplasts that allow them to capture energy by basking in the sun, yet they can also be voracious predators, slurping up other cells with a personal feeding tube. Some dinoflagellates produce deadly toxins that can result in massive fish kills, such as red tides. Luminescence in dinoflagellates originates from microsomes, called scintillons, scattered throughout their single cell. They are responsible for the sparkle visible in a boat's wake and the flashes of light that accompany waves on a moonless beach night. Dinoflagellates mainly sparkle when they are disturbed, the bioluminescence acting as a predator alarm. Along with other glowing marine organisms, dinoflagellates transform the dark ocean into a bioluminescent minefield. Any movement could set off a flash, revealing the location of the animal to lurking predators.

Certain bodies of water contain extremely high concentrations of bioluminescent dinoflagellates. A bioluminescent bay in Vieques, Puerto Rico, contains almost 6,000 dinoflagellates in each tablespoon of seawater, hundreds of times more than in the open ocean. These conditions produce brilliant light shows surrounding swimmers in the bay. Charles Darwin, while aboard the HMS *Beagle*, witnessed a spectacular dinoflagellate display: "While sailing a little south of the Plata on one very dark night, the sea presented a wonderful and most beautiful spectacle. There

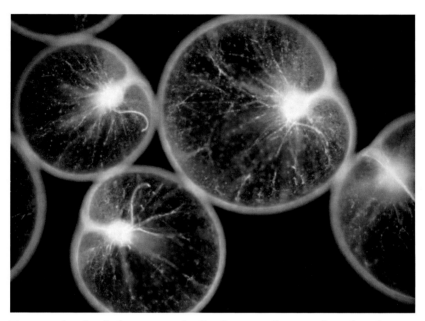

The most abundant bioluminescent dinoflagellate, *Noctiluca scintillans*. Photo by Wim van Egmond.

was a fresh breeze, and every part of the surface, which during the day is seen as foam, now glowed with a pale light. The vessel drove before her bows two billows of liquid phosphorus, and in her wake she was followed by a milky train. As far as the eye reached, the crest of every wave was bright, and the sky above the horizon, from the reflected glare of these livid flames, was not so utterly obscure as over the vault of the heavens."[14]

As Darwin noted, a ship moving through the ocean at night churns the water and produces a sparkling light show in its wake. However striking such displays are to witness, naval personnel are not fond of them. The bioluminescence produced as vessels move at night causes several logistical problems for naval operations. During World War I the commander of the German U-boat *Deutschland* described a night when the luminescence surrounding the boat was so intense as to make identification of objects on the horizon nearly impossible: "The phosphorescence of the sea seriously hindered the lookout. One was almost blinded, the eyes grew painful, and the vision became unsteady through the persistent coruscation of the waves in the coal-black night. This was rather uncomfortable, for we had now reached a region that was intersected by many steamer routes, and it was necessary to take double precautions."[15]

Following World War II, the Russian navy secretly studied marine bioluminescence to determine how it affected its operations. A prominent Russian naval officer, Nikolai Ivanovich Tarasov, who headed the investigations, wrote in 1956 about the disturbing effects of nighttime biolumi-

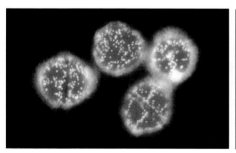 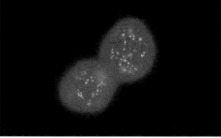

Bioluminescent dinoflagellates *(Lingulodinium polyedrum)* during the night *(left)* and during the day *(right)*. The white/blue spots are organelles called scintillons and are where light emission occurs. The red fluorescence is chlorophyll. Photos by J. Woodland Hastings.

nescence on navigation at sea: "The luminescence of a section of the water surrounding a vessel impedes the observation of the rest of the sea surface, the visible horizon, the air and shores. Indeed, from a ship moving in brightly luminescent waters, it is hard to distinguish meagerly lit distant areas, which, owing to the sharp contrast, are shrouded in darkness. Competition between the luminescence of the water and the lights of distant ships and coastal stations is frequently the cause of misconception. By intensifying the background light, luminescence of the sea diminishes the distance of visibility."[16]

Submarine crews often use the luminescent trails to track the path of torpedoes as they approach their targets. But bioluminescence can also reveal the location of vessels to enemies. On the night of November 9, 1918, in the Mediterranean Sea near Gibraltar, Spain, a British Q-ship noticed a large glowing shape beneath the surface. The Q-ship launched three 76-millimeter missiles and discharged a series of depth bombs. The large glowing mass was the German submarine U-34. "The phosphorescence of the sea was so intensive that the movements of the shining U-34 under the surface were clearly visible," Tarasov wrote.[17] Within 30 minutes of being located, U-34 was destroyed—the last German submarine to be sunk in World War I.

Carrier-based aviators in World War II often used the long bioluminescent trail churned up by ships' wakes to pinpoint vessels at night. One particularly famous incident is captured in the book *Lost Moon* (and recounted in the movie *Apollo 13*). James A. Lovell, one of the astronauts who made the ill-fated Apollo 13 mission to the moon, was once saved by bioluminescent dinoflagellates. In February 1954 when he was a navy pilot, Lovell set out on a night training mission from a carrier off the coast of Japan. While he was taking off in stormy weather from the USS *Shangri-La*, his directional finder malfunctioned, heading him in the wrong direction. To make matters worse, his instrument panel suddenly short-circuited, burning out all the lights in the cockpit:

> Lovell's heart went timpanic. His mouth went dry. He looked around himself and could see absolutely nothing; the blackness outside the plane had suddenly come inside. Tearing off his oxygen mask, he gulped a breath or two of cabin air and thrust a penlight in his mouth to shine off the instruments. The silver-dollar-sized beam the tiny flashlight produced danced cross the dashboard, dimly illuminating one needle or dial at a time. Lovell checked the readings as best as he could and then fell against the seat to consider what he should do next.
>
> Lovell took the penlight out of his mouth, switched it off, and scanned the darkness. Down below him at about two o'clock, he thought he noticed a faint greenish glow forming a shimmery trail in the black water. The eerie radiance was barely visible and would have been lost to Lovell altogether had the blackness in the cockpit not acclimated his eyes to the darkness. But the sight of it made his heart leap. He was certain he knew what the strange radiance was: a cloud of phosphorescent algae churned into luminosity by the screws of a cruising carrier. Pilots knew that the spinning propeller could light up organisms in the water, and this could help them locate a missing ship. It was one of the least reliable and most desperate methods of bringing a lost plane home safely, but when all else failed, it could sometimes do the trick. Lovell told himself that all else had indeed failed, and with a fatalistic shrug he peeled off in pursuit of the dim green streak.[18]

After such experiences the navy soon recognized the strategic need to predict the distribution of planktonic marine bioluminescence in wartime, not only to use it as Lovell had, but to enable captains to avoid bioluminescence and thus disguise the whereabouts of their vessels. As sophisticated satellite imaging became available during the Cold War, bioluminescent blooms could easily give away the presence of nuclear-powered submarines, whose strategic advantage lay in their ability to remain underwater for months at a time without surfacing, hiding their location. Satellites can detect the amount of light given off by a single match at the ocean surface. The navy began an aggressive program to map and possibly neutralize marine bioluminescence so their nighttime maritime operations would not be compromised. But predicting when and where bioluminescent animals congregate is difficult. In one incident during the early phases of the 1991 Persian Gulf War, a team of Navy SEALs surreptitiously swam onto a Kuwaiti beach pulling an egress line, used to provide a rapid escape route. After landing in enemy territory, the SEALs looked back at the water. They were shocked to see their line glowing blue. Millions of bioluminescent dinoflagellates, moving with the swift current, were striking the line, each releasing a pulse of blue light. One of the SEALs later recalled that had there been an alert enemy on shore, the team would have been easy targets.[19]

SEALs also use an underwater delivery vehicle (SDV) designed to covertly transport teams underwater to hostile targets. In planning their nighttime operations with SDVs, the SEALs take account of seven "Critical Meteorology and Oceanography Thresholds": water current, wave height, tides, water quality and temperature, lunar illumination, and bioluminescence. The SEALs don't use their underwater vehicles when bioluminescence causes "visible detection of an SDV submerged 10 feet under ambient light."[20]

So important to the navy is the detection of bioluminescence that the Office of Naval Research funds a great deal of the work of the Bioluminescence Department at the Harbor Branch Marine Oceanographic Institution, in Fort Pierce, Florida. The department's director, Edith Widder, is a leading expert in devising instruments to detect, monitor, and quantify marine bioluminescence. Her inventions not only quantify marine bioluminescence but also determine which species are responsible for the light. The navy uses such devices in advance of nighttime naval operations to predict the bioluminescence liability.

Widder's devices monitor the bioluminescence from all glowing organisms, including the larger jellyfish and salps. Another investigator, Mark Moline, a bioluminescence researcher at California Polytechnic State University, is focusing on the smaller animals. In 2001, Moline was attending a navy conference on bioluminescence when he heard a navy lieutenant and former SEAL detail his encounter with glowing dinoflagellates in Kuwait. "I saw the concern in his eyes and from that point on, approached this problem with more energy and with a focus grounded in science," says Moline. At his laboratory in San Luis Obispo, Moline is currently designing remotely operated underwater rovers to predict the near-shore bioluminescence, mainly from dinoflagellates. These torpedo-shaped propeller-driven crafts glide silently beneath the surface and sensitively measure bioluminescence in shallow-surf conditions.

CHAPTER TWO

Fireflies of the Sea

THE STUDIES Raphaël Dubois conducted laid the foundation of bioluminescence research, but the man who is most associated with popularizing the study of living light is Edmund Newton Harvey. As Frank Johnson, a prominent bioluminescent researcher and former student of Harvey, put it: "It has been aptly remarked that no man makes an institution, but if that is true, Edmund Newton Harvey during his lifetime came very close to establishing an exception to the rule . . . No one has ever matched, and perhaps no one ever will, his widely encompassing, scholarly contributions to our knowledge of the mystifying natural phenomenon of the emission of visible light by living organisms."[1]

Harvey was born on November 25, 1887, in Germantown, Pennsylvania, a streetcar suburb of Philadelphia, where he lived in a large stone house. His father was a minister who died when Harvey was six years old, and he was raised by his mother and three older sisters. As a child, Harvey displayed a fondness for living organisms and their systematics, often digging for insects on the several acres of land surrounding his house. He was known to collect "every conceivable natural object" and display the skeletons prominently in his room. His family nurtured his fascination for natural history by allowing him to keep frogs in the family bathtub to lay eggs in the spring. He viewed the natural world as his temple and while in church on Sundays he recalled his "restlessness" and desire "to be outside collecting things."[2]

After graduating from Germantown Academy in 1905, Harvey entered the nearby University of Pennsylvania. He became so enthralled with scientific pursuits that he participated in virtually no social activities beyond associations in class or laboratory. He later recalled: "As I was only interested in science, I felt no need for anything else."[3] In college, Harvey began to study cellular morphology and biochemistry. Any additional time he spent at the nearby Academy of Natural Sciences examining some of the 13,000 species of many-legged arthropods in the class of animals called Myriapoda. This led to his first ambition: to become an international authority on centipedes. Harvey complained of obligations at the university that kept him indoors: "Nothing can be more ridiculous than to go to the stuffy, smelly dressing room, put on some dirty, smelly gym clothes, then ascend to a large bare hard-floored room, not even well ventilated and push up dumbbells at the behest of an instructor. I was fundamentally an outdoors man who spent long hours tramping through woods and over hills."[4]

In September 1909, Harvey moved to New York City to begin his doctoral research at Columbia University with Thomas Hunt Morgan. At that time, Morgan was an embryologist beginning to expand his research into the genetics of the fruit fly *Drosophila melanogaster.* The year Harvey arrived at his lab, Morgan observed a discrete variation in a single male fly: it had a white eye. His curiosity aroused, Morgan bred the white-eye fly with the normal, red-eyed females. All of the offspring were red-eyed. Matings among this generation produced a second generation with some white-eyed flies, all of which were males. To explain this curious phenomenon, Morgan developed the hypothesis of sex-linked characteristics, which he postulated were part of the X chromosome. Morgan's work, which won him the Nobel Prize in 1933, is recognized as a cornerstone of modern genetics. His chromosomal theory of heredity changed the face of biology, and his laboratory launched the careers of some of the most influential classical geneticists in the United States, including Hermann Muller, who won the Nobel Prize 13 years after Morgan for his discovery that X-rays cause mutations in *Drosophila*.

When Harvey arrived at Columbia, Morgan's interests spanned both embryology and genetics. Harvey found such a diverse set of interests daunting:

> Morgan's lectures were sometimes supplemented by laboratory experiments or demonstration . . . I recall that one day we had a tubful of frogs for experiments on behavior and reaction to light. These frogs were supposed to do something in relation to the direction of a light beam, but I noticed that on this day all the frogs were recalcitrant. When placed in the beam they either did not move, or if prodded, jumped in the wrong direction. I decided that the study of animal behavior was one involving too many unknown variables, not the subject for me; that I had better stick to the single cell, which is complicated enough, but cannot compare with an adult vertebrate animal.[5]

These comments reflect the reductionist philosophy that grew in popularity in the twentieth century. As the complexity of life became more apparent, it was recognized that before you could understand what was happening to the system as a whole, you needed to comprehend its smaller, building-block components. This meant that it was necessary to understand what was happening on a cellular level before understanding processes like behavior. With this in mind, Harvey committed himself to study the permeability of cells, a field vastly different from Morgan's. Within two years, Harvey completed a Ph.D. thesis detailing the characteristics of the cell membrane.[6] Immediately thereafter, he was offered a position as instructor in the expanding Department of Biology at Princeton University. Only 23 years old at the time, he was often mistaken for an undergraduate.

At first, Harvey found the bucolic college town and serene atmosphere of Princeton University less than stimulating. In the early 1900s Princeton was fundamentally concerned with the humanities, a topic of no particular interest to Harvey because the "philosophical approach was not definite enough." Yet, as Harvey put it, he was "determined to stick it out" in this new setting.[7] Lonely, pining for his companions and the rhythm of New York, he joined the Princeton Bachelors Club, a group formed primarily by preceptors in the various disciplines of the humanities. There he found himself attending evening lectures on subjects that he said he would never have attended in the past. Although he became exposed to a world of new ideas, he remained apolitical: "My mother taught me strict honesty in all things and a particular dislike of insincerity and hypocrisy, which probably explains my lack of interest in political affairs."[8]

At Princeton, Harvey continued his focus on the structure and permeability of cell membranes. In the 1910s, the structure of a cell's outer plasma membrane was largely unknown. Harvey's membrane structure research culminated with the publication of a landmark study with a researcher in his laboratory, James Frederick Danielli. The model proposed, in this and in subsequent papers by Danielli, is widely regarded as the first accurate description of the cell membrane.[9] It was understood at the time that all cells were bound in a protective bubble, separating the inner components of the cell from the harsh outside world. It was known that this outer cell membrane was an oily substance. Harvey and Danielli determined that cell membranes are composed of only two

Bioluminescent bacteria *(Vibrio harveyi)* growing on agar plates. Photo by J. Woodland Hastings.

sheets of soapy molecules stacked upon each other like pancakes, known as the lipid bilayer. Each pancake has two different faces—one that is watery and one that is oily. When the two pancakes come together, the oily sides face each other on the inside, while the wet sides face outward. Danielli would later propose that membrane proteins exist, embedded in the bilayer like giant blueberries. These membrane proteins act as selective portals, regulating what enters and exits a cell. The lipid bilayer is one of the foundations of modern cell biology.

In the spirit of nineteenth-century naturalists, Harvey would take extended "voyages of discovery" to exotic places to study and catalog strange organisms, often writing the first description of them, incorporating the creatures into the lexicon of science. As a testament to his influence, several organisms were later named after him, including a centipede *(Pselloides harveyi)*, luminescent bacteria (*Achromobacter harveyi* and *Vibrio harveyi*), and a firefly *(Photinus harveyi)*.

On a series of expeditions that rival Darwin's accomplishments on the *Beagle* voyage, Harvey would visit American Samoa, Hawaii, Cuba, Japan, Korea, Manchuria, the Philippines, Singapore, Batavia, Semarang, Surabaya, Bali, Lombok, Sumbawa, Macassar, Amboina, and the Banda Islands. In 1913, Harvey set out on an excursion that would forever change his scientific life. Along with Alfred Mayor, a former professor from the University of Pennsylvania, he made a trip to the South Pacific, visiting Australia's Great Barrier Reef, Tahiti, Raratonga, Wellington, Sydney, Brisbane, Townsville, Cairns, and the Thursday and Murray Islands. It's not clear exactly when on this trip he became smitten by bioluminescence. His first paper on bioluminescence, entitled "On the Chemical Nature of the Luminous Material of the Firefly," appeared in 1913.[10] But it *is* clear he was hooked for life during his honeymoon in Japan in 1916. While swimming at night in the waters outside the Misaki Laboratory Biological Station about 40 miles south of Tokyo, he was seduced by a little glowing crustacean, *Cypridina hilgendorfii*. Abundant in the shallow waters of the Sea of Japan, this animal is known locally as Umihotaru, or sea firefly. These voracious scavengers scurry along the ocean bottom, waiting for animals to perish and sink. A small group of *C. hilgendorfii* can consume an entire fish in a matter of hours. They also actively hunt shrimp many times their size. Fortunately, his young bride, Ethel Nicholson Browne, was sympathetic to his preoccupation with marine creatures. Only three years before their wedding she had completed a doctorate in biology from Columbia University, studying the male gametes of an aquatic insect. She later devoted her career to the embryology of sea urchins.[11]

Before leaving Japan, Harvey made arrangements to collect, dry, and ship large quantities of the glowing crustaceans back to Princeton. He considered *Cypridina* "to be by far the best [organism] for the biochemical investigation of luminescence" because the dried animal sprang into brilliant luminescence when moistened with water, even after years of storage.[12]

Cypridina, a cousin to shrimp and crabs, is the size of a sesame seed

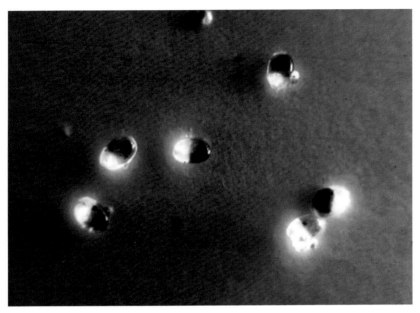

The bioluminescent ostracod *Cypridina*. Photo by Toshio Goto.

with two hinged plates of body armor. Active mainly at night, these animals emit a dim glow of blue light when they are undisturbed. But when pursued, they dart away and release plumes of bright blue light, creating large diffuse clouds of bioluminescence. The glowing excretion hangs in the water for several seconds and acts as a subterfuge, confusing the pursuing predator and allowing *Cypridina* to escape.[13] Male sea fireflies also secrete small glowing plumes to attract mates. At dusk on reefs around Japan, female *Cypridina* congregate on the shallow bottom for a nightly light show. Male *Cypridina* then make straight and rapid ascents from the bottom to the surface of the water. During the ascent they squirt out small plumes of a bright blue material in a punctuated pattern. The luminescent material remains bright for several seconds and hangs in the water column. The result is an accumulation of luminous trails that resemble strings of glowing pearls. With hundreds of males participating, the water comes alive with a brilliant spectacle. The spacing of the "pearls" is species-specific; starstruck females connect the dots and find a waiting mate.

In the early 1940s, the Japanese military developed a scheme to use *Cypridina* as a tool on the battlefields of New Guinea and other sites in the Pacific theater.[14] During long marches on moonless nights in the jungles of the South Pacific, Japanese soldiers had difficulty seeing each other since battery-powered lights could easily disclose their location to the enemy. The plan involved distributing small vials of dried sea fireflies to the troops. Each soldier was to rub small amounts of the bioluminescent animal onto the back of the soldier in front of him; this would allow soldiers to navigate pitch-black trails while still keeping 15 to 20 feet apart. The military also planned to use the *Cypridina* to read maps, relying on the dim bioluminescent glow in circumstances when the glare of a flashlight would compromise safety. "A soldier," Frank Johnson wrote, "is claimed to have spit on a little of this powder in his hand and thereby gotten enough light to read a map, with no likelihood of arousing suspicion."[15] During World War II, hundreds of pounds of *Cypridina* were collected by Japanese officers, students, and volunteers. Some of the collection took place in Tateyama, near the Misaki Marine Laboratory, where Harvey had collected a few decades earlier. The technique developed for Harvey was replicated for the military collecting. A large fish head was tied to a string and dropped to the shallow, sandy bottom. *Cypridina* infested the fish head within two hours, while feeding on the flesh. When the decaying head was hauled to the surface, Harvey wrote, the glowing *Cypridina* were "easily picked off."[16] The animals were then placed in the sun to dry before being sent to the troops. It is unclear if the sea firefly was actually used in the war, because much of the dried powder was lost in transport when U.S. submarines sank Japanese carrier ships. It is also reported that some quickly became useless in the high humidity of the tropical climate.[17]

When Harvey returned from his honeymoon, he found that Raphaël Dubois, now 68 years old, had shipped him a jar of *Pholas* siphons preserved in sugar from his laboratory at Tamaris-sur-Mer, France.[18] Harvey did not realize it at the time, but he would spend the majority of his career confirming and extending Dubois's work on bioluminescence. As Frank Johnson wrote, Harvey "indefatigably sought evidence for the existence of a 'luciferin-luciferase system' in virtually every type of luminescent organism he could get his hands on."[19]

From a biochemical point of view, bioluminescence offered an attractive subject to scientists in the early 1900s. Without expensive equipment, it is possible to mix two reagents together (containing hot and cold extracts) and measure the kinetics of a chemical reaction simply by quantifying the amount of light given off. Harvey soon found components analogous to luciferin and luciferase in several types of luminescent organisms: the American fireflies (*Photinus* and *Photuris*), the Japanese firefly *(Luciola)*, the Bermuda fireworm *(Odontosyllis)*, and *Cypridina* (the sea firefly). Harvey also hoped to prove that luciferin and luciferase from different species were interchangeable, showing that all bioluminescent animals evolved from a common ancestor. He had early success supporting this idea by mixing luciferin and luciferase of closely related firefly species and discovering that they still ignite. This finding inspired

Harvey, only 28 years old, to pen a piece published in *Science* in which he declared confidently, "In a general way, then, we may say that the problem of bioluminescence has been solved at least in its broad aspects. There still remain many details to be filled in, details which will take some time to complete."[20] The first part of this statement turned out to be grossly naive, while the second half proved prophetic. Many details of bioluminescence do still await discovery.

Over the next 30 years, Harvey found that luciferins and luciferases from different animals were noninterchangeable, which suggested that bioluminescence evolved separately in several different species and, therefore, serves a variety of different functions. The discovery of major compositional differences in luciferase and luciferins further demonstrated the vast differences in the bioluminescent apparatuses of different animals. These revelations led Harvey to state much later in life: "No clear development of luminosity along evolutionary lines is to be detected but rather a cropping up of bioluminescence here and there, as if a handful of damp sand has been cast over the names of various groups on a blackboard, with luminous species appearing wherever a mass of sand stuck."[21]

During his career, Harvey returned again and again to *Cypridina*. In the 1950s he sought to determine the chemical nature of the *Cypridina* luciferin/luciferase reactants. What kind of a chemical reaction can a living organism perform that produces light but not heat? All that was known is that the luciferase is the catalyst and the luciferin is the fuel. To determine their chemical nature, Harvey first needed to purify each component. He enlisted several prominent organic chemists— Rupert S. Anderson, Aurin Chase, Howard Mason, and Fred Tsuji—to join him at Princeton to develop procedures to concentrate and purify *Cypridina* luciferin. This turned out to be a difficult task for Harvey's newly assembled group. Although dried animals produce a deceivingly bright light when ground up and mixed with water, each animal contains only a minute quantity of both the luciferin and the luciferase, only about a millionth of its weight. Moreover, luciferin is highly unstable in aqueous solutions containing dissolved oxygen.

In 1935, the Princeton group had devised a method of partially purifying *Cypridina* luciferin.[22] This method resulted in an extract with 2,000 times the luciferin activity by weight when compared to that of the dried animals. By the 1950s, however, biochemists wanted more than a concentrated sample—they wanted a solution pure enough to form crystals. When a pure solution of a particular compound is concentrated above a critical threshold, the molecules will associate in an ordered lattice and leave the solution in pure crystalline arrays. Harvey explained: "Many of the chemical tests applied to crude solutions to determine the nature of luciferin and luciferase are of little value, and later work with partially purified material has indicated that luciferin and luciferase in crude *Cypridina* extracts behave very differently from purified material."[23] Although the Princeton chemists had produced a concentrated solution of luciferin, it was not pure enough to obtain crystals. Over 40 years the Harvey laboratory tried without success to concentrate the bioluminescence components to purity.

Toward the end of his 45 years at Princeton, in the late 1950s, Harvey became disillusioned by his inability to purify these components and take the chemistry of bioluminescence to the next natural step. The basic question of how bioluminescence works had not been solved. Harvey could not isolate and identify the true chemical nature of any luciferase or luciferin, and so it remained unclear whether these constituents were proteins or some other type of molecule. In the last years of his scientific career, perhaps out of frustration, he focused his attention on a meticu-

Edmund Newton Harvey and his wife Ethel Nicholson Browne Harvey at Princeton University. Courtesy of Princeton University Library, Princeton, New Jersey.

lously detailed 692-page opus, *A History of Luminescence from the Earliest Times until 1900*, obsessively detailing literary mentions of luminescence.[24] Frank Johnson, a fellow professor of biology at Princeton stated, "Final purification has been one of the chief objectives, for its own sake, since the days of Dubois. At this point, it was more than ever a crying necessity."[25] Without purified constituents, the kinetics and nature of the light-producing reaction could not be studied.

But help was on the way. In 1957, the full purification of *Cypridina* luciferin was achieved by an unknown Japanese scientist working in isolation. Describing the importance of this discovery, Johnson wrote: "The stage was now set for definitive advances. Moreover, for the first time in nearly 300 years (since Robert Boyle, who was vastly ahead of his time in the seventeenth century), there was now on the scene a dedicated scientist, gifted with a rare instinct for sensing the right solutions to perplexing problems and devoting his inexhaustible energy and skilled efforts to bioluminescence. His name is Osamu Shimomura."[26]

CHAPTER THREE

From the Fires of Nagasaki

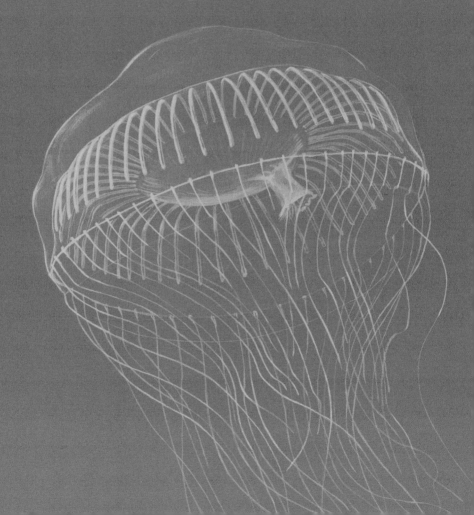

AS THE BRAND-NEW blue station wagon purred along U.S. Highway 2 in northern Montana, Osamu Shimomura sat back and peered at the expansive mountains of Glacier National Park. It was the summer of 1961, and Shimomura, a young Japanese Fulbright Scholar and biochemist, was happy to be on the road. He had embarked a few days earlier from Princeton University on a 3,000-mile cross-country road trip to Puget Sound in Washington state. At the wheel was his boss, Frank Johnson, a portly Princeton biology professor who spoke fluent Japanese with a distinctive North Carolinian accent. Also in the car was Shimomura's wife, Akemi, who had arrived from Japan only a few days earlier. Shimomura had spent the previous year living alone in a sparsely furnished apartment on Princeton's Nassau Street. One of the few decorations in his house was a hand-drawn rendition of the chemical structure of *Cypridina* luciferin, taped to his front door. Though his accommodations were modest, that was not evident in his attire. Shimomura always dressed neatly. Even on the road trip—while spending 12 hours each day in the car—he wore a button-down shirt, a tie, and a cardigan.

The goal of the journey was simple: to unravel the mystery of how a palm-sized jellyfish glowed. When Shimomura returned to Princeton on the same stretch of road a few months later, not only would he have solved one of the greatest puzzles of bioluminescence, but he would also have discovered a strange fluorescent substance that would, eventually, light up the entire field of biology.

Shimomura came of age during one of Japan's most difficult eras. Growing up as the son of an army colonel during World War II, he lived a nomadic life, moving from Sasebo to Manchuria to Osaka and finally, in July 1944, to Isahaya, a quiet farming village on the outskirts of Nagasaki. Shimomura was 15 years old when his family settled in Isahaya. His father, stationed in Thailand, had realized that Japan was slowly losing the war and had ordered his family away from Osaka because he feared that it would be the target of firebombing by the increasingly aggressive U.S. military. Shimomura, his mother, and his grandparents moved to a quiet countryside cottage overlooking Mount Tara, 10 kilometers from downtown Nagasaki.

On his first day at Isahaya High School in September 1944, Shimomura and his classmates were informed that there would be no classes because students were needed to assist in the industrial war effort, a common practice in Japan during World War II. Half of the school's 300 students were sent to work in a naval airplane factory in Ohmura while the other half went to a shipbuilding factory in Nagasaki. Shimomura was assigned to Ohmura, about 100 kilometers from Nagasaki. He slept in a cramped 10-foot-square dormitory with six other friends. The meals were meager and low in nutrients: typically a bowl of cooked rice, wheat, and defatted soybeans, a mixture normally used to feed cattle. Occasionally, there was a cup of miso soup, slices of pickles, and a dish of fish or vegetables. Even now, Shimomura remembers constant hunger.[1]

Less than a month after Shimomura began working in Ohmura, the United States targeted his factory with about 100 B-29 bombers armed with high explosives. As warning sirens wailed, Shimomura ran to leave the building with his classmates. He stopped briefly on the way out to peer up at the sky, mesmerized by the spectacular view of shiny planes flying in tight geometric formation. His awe soon turned to horror as the planes unleashed a maelstrom of fire and explosions, destroying the factory and killing several of his classmates. The bombing stopped nearly as quickly as it began, and the remaining students regrouped. Under orders from their supervisors, the students returned to a burning hangar in a feeble attempt to rescue fighter planes trapped among the wreckage. A second wave of bombers approached while they were pushing a crippled plane through the flames and began to rain down magnesium bombs. First unleashed on Dresden, Germany, these bombs inflamed buildings by producing intense flashes of heat. When the second wave of bombing began, Shimomura abandoned the planes and again scrambled for his life, zigzagging through the raging flames before finding safety in the underbrush alongside a nearby airstrip. Many more of his classmates were again unable to escape. With the factory no longer operational, Shimomura was sent back to his grandparents' house in Isahaya, where he tried to forget what he called "the most miserable period" of his life.

In early 1945 the U.S. Army Air Corps devastated most of Japan's major cities, killing an estimated 330,000 people. Shimomura's father had shown foresight in moving his family from Osaka because the fire-bombing on the night of March 13, 1945, turned an area of 25 square miles into, as one survivor recalled, "a smoldering desert."[2] In all, Allied planes dropped 160,800 tons of bombs on the home islands of Japan. The vast majority, 153,620 tons, was dropped between March 9 and August 15, 1945.[3]

Shortly after Shimomura's return to Isahaya, the Japanese navy built a less conspicuous airplane repair factory, consisting of ten wooden buildings dispersed in the foothills around Nagasaki. Shimomura was reassigned there to repair engine casings and mend parts of crippled fighter planes. He enjoyed this job much more than his work in Ohmura, because the factory was only three miles from his home and he could therefore walk to work each day on a beautiful country road surrounded by the sound of chirping cicadas. "Our work load was heavy at first then rapidly decreased, reflecting the consumption of airplanes in the war, probably in kamikaze attacks," he recalled.

The morning of August 9, 1945, began as a typical hot and humid day in Isahaya. Shimomura, now 16 years old, arrived at work dressed in shorts, a crisp white shirt, and sneakers. At 10:57 A.M. the student workers heard the familiar whine of air raid sirens signaling the approach of enemy bombers. "We went out the building and climbed a nearby hill, rather than going into an underground bunker; it was against the rules, but we knew from our experience of many air raids that we would be safe there." Hands cupped above his eyes, he squinted at the pale blue sky. Expecting to see a squadron of American bombers, he was relieved to see only one plane dotting the horizon, heading southbound toward Nagasaki, 12 kilometers away. The American bomber passed by at high altitude and dropped only three white parachutes that wobbled slowly toward the ground. The parachutes were puzzling since they did not appear to carry paratroopers. Shimomura and his fellow workers heard a few scattered gunshots, probably aimed at the parachutes. Moments later another bomber passed above, heading in the same direction. Relieved that the planes did not seem to pose a considerable threat, Shimomura ran down the hill and returned to work. He recalls the next horrifying series of events: "At the moment I started to work, a bright flash filled the interior of the building. It was so bright that my eyes were temporarily blinded. Less than a minute later came a loud thunder of explosion and a strong pressure wave that was painful to my ears. I noticed the sky was quickly covered by cloud. All these things seemed mysterious."

By the late afternoon he had learned that a massive explosion had taken place in nearby Nagasaki. He finished work early and began his walk back to his grandparents' house. The day that had started out with clear skies had turned "strangely overcast." During his homeward journey, a black rain began to fall upon the surrounding countryside, leaving an eerie black coating. The raindrops stained his white shirt, and by the time he arrived home, it was charcoal gray. His grandmother, alarmed by the billowing dark cloud enveloping the mountainous countryside, immediately instructed Shimomura to remove his soiled clothes. "I washed my body and changed clothes although I was not aware of the rain's dangerous radiation," Shimomura remembers. "The radio reported in the evening that Nagasaki was attacked with a special bomb, similar to the one that exploded in Hiroshima three days earlier, and that [the] Urakami district was heavily damaged."

When Shimomura arrived at work the next morning, the chief officer of Isahaya's airplane repair facility ordered all workers into pickup trucks to help people in Nagasaki. A port city surrounded by mountains, Nagasaki was accessible only by two routes in 1945: a single-track railway and a highway that went through a tunnel at Himi Pass and reached the east end of the city. On that day, however, the tunnel was impassable, so they followed an obscure back road toward Urakami, finally reaching a place where houses were still smoldering. Shimomura recalls: "We were, however, suddenly ordered to go back without explanation: the directing officer might have found that the road was impassable further on, or he might have sensed a strong radiation, or he might have thought that young boys would not be helpful under the circumstances."

The thermal pulse from the atomic bomb dropped on Nagasaki was so intense that it imprinted negative-image shadows on buildings and streets. A man pulling a handcart left his silhouette on the pavement before being carbonized; the dark patterns in women's blouses and kimonos seared tattoos in their flesh while the white fabric reflected the heat.[4] The parachutes that Shimomura saw falling from the sky carried devices that measured the intensity of the explosion and transmitted various data back to the plane that dropped them, recording the equivalent of 22 kilotons of TNT, nearly twice the intensity recorded over Hiroshima three days earlier.[5] The bright flash that momentarily interrupted Shimomura's work had instantly ended 40,000 lives and injured another 40,000 people,[6] resulting in the unconditional surrender of the Japanese military and the end of World War II six days later.

At war's end, Shimomura's duties at the factory ended, and Japan was plunged into chaos. Looking for guidance, Shimomura returned to his high school every 2 to 3 days hoping to find instructions for students, but the school was being used as a temporary hospital for those fleeing Nagasaki. "It was filled with hundreds of people, mostly injured and burned, whose names were displayed on large white sheets of paper at the front gate," Shimomura recalls. "Many of the names were crossed out every day, indicating that the person had been taken away by a relative or that the person had died." Two weeks after the blast, Shimomura returned one hot and sunny afternoon, again hoping to find instructions for students. When he approached the front gate, he saw that more than half the names were crossed off. A few people were silently loading dead bodies onto a cart outside. Shimomura was distressed to see the bodies covered only by fresh straw mats that exposed the feet of the dead, make-shift coverings necessitated by the shortage of coffins. He walked farther inside the schools gates. "On the left side was a large open ground where several people were strolling, very slowly, step by step, under bright sun-light." Shimomura walked closer and saw that they had black medica-tion that looked like coal tar covering their burns. He approached a "stroller" whose back was totally black, except for a few intermittent white specks. "It took a minute to realize that those were maggots . . . that had hatched on the human flesh." When he looked around, he real-ized everyone had maggots burrowing into their dead flesh. "Apparently, the strollers were spiritually dead. I thought I was looking at ghosts, in broad daylight. I was shocked. My heart was frozen, my brain became unfeeling and the sound of cicadas stopped. That quiet scene was im-printed in my memory much stronger than any other bloody or grue-some scene I had ever seen."

During a series of interviews conducted over a two-year period, ending in 2004, in his spartan study in Woods Hole, Massachu-setts, Shimomura stared out the window and vividly recalled these events from nearly 60 years ago. He has spoken very little about the cata-clysm that he witnessed. When asked what he did in the years immedi-ately following the war, he replied, "There was no choice in our future." "We just had to live." Japan's infrastructure was badly damaged, his school records were destroyed, and most of his teachers were killed in the war. In fact, Shimomura's high school education was sacrificed to the war effort; his studies were not resumed, and graduation was not an op-tion. After the war, all his university applications were rejected.

Six hundred out of the 850 medical students at the Nagasaki Medical College were killed and most of the others were injured, and of the 20 faculty members, 12 were killed and four others injured.[7] The College of Pharmaceutical Sciences relocated to an old military base in Isahaya. Without any experience in pharmacology, Shimomura applied and was finally accepted to the makeshift university in 1948. With most professors killed in the nuclear blast, inexperienced instructors taught his classes, and Shimomura acquired most of his knowledge through independent study. Three years later Shimomura graduated and applied for employment at Takeda Chemical Industries, the largest pharmaceutical company in Japan. The interviewer deemed him unfit for industry work and turned him down.

Luckily, Shimomura's former instructor of analytical chemistry at the College of Pharmaceutical Sciences, Shungo Yasunaga, offered him a position as a teaching assistant. Yasunaga was in the process of completing his doctorate from Kyoto University while also teaching at the University of Nagasaki. Besides carrying out his teaching duties, Shimomura helped Yasunaga with many aspects of his thesis on separation chemistry. Yasunaga's work in the early 1950s involved the development of various laboratory methods to separate mixtures of small molecules. Such separation techniques allowed the purification of small molecules from cells or the products of chemical reactions used in making pharmaceuticals. Shimomura and Yasunaga coauthored eight papers on the chromatographic separations of small molecules, all in the *Yakugaku Zasshi*, the journal of the Pharmaceutical Society of Japan. After four years, Yasunaga, impressed by Shimomura's hard work and dedication, felt honorbound to reward Shimomura for helping him with his thesis. Yasunaga accompanied Shimomura to Nagoya University in 1955, in hopes of introducing him to the noted biochemist Fujio Egami, and possibly even of securing a position for him.

When they arrived at Nagoya University, they were disappointed to learn that Egami was out of town attending a meeting. Even years after the end of the war, the phone lines were not fully operational, and communication remained difficult. While surveying the chemistry department, they bumped into the newly appointed 40-year-old professor Yoshimasa Hirata. Hirata had received his doctorate only six years earlier from Nagoya University. Sometimes mistaken for a student because of his disheveled appearance, Hirata enjoyed doing laboratory research aimed at isolating and purifying natural compounds.

Yasunaga explained to him how they had made the long overnight train journey to introduce Shimomura to Egami. To their bewilderment Hirata replied: "Sure, you can come to my lab anytime," and then hurried back to his laboratory, leaving Yasunaga and Shimomura staring at each other in confusion. Hirata had difficulty hearing in one ear: he thought that they said they had traveled to Nagoya to meet with him. Those who worked closely with Hirata knew always to speak loudly into his good ear if they wanted to be understood.

"What are you going to do?" asked Yasunaga, visibly upset as they reached the building exit. "I don't care," replied Shimomura as they began walking toward the train station. "I'll work with anybody." So one month later (and while he was still on the payroll at Nagasaki's College of Pharmaceutical Sciences), Shimomura showed up at Hirata's laboratory to begin work as a visiting researcher. On Shimomura's first day, Hirata pulled down a large desiccators of dried *Cypridina*, took out a pinch, crushed the pieces in his palm, and added water. Instantly the material began to glow blue. "We know nothing about this," he said, referring to the chemical properties of the light reaction. "Just that it glows." Hirata explained that he would never assign a graduate student to the problem of *Cypridina* bioluminescence since there was an overwhelming chance of failure. He was aware that Princeton University scientists, led by Harvey, had failed in a 40-year attempt to purify the components of the reaction. But since Shimomura was only a visiting researcher, Hirata thought there was nothing to lose by setting him to work with the abundant material.

Hirata asked Shimomura to isolate and study *Cypridina*'s luciferin, the chemical fuel that causes its bioluminescence. Absolutely pure luciferin was needed in order to determine its chemical structure. Without any assistance other than the available literature (mainly in English), Shimomura set about his task. When Shimomura began his work with Hirata, he knew that luciferin was more abundant than the catalyst, the luciferase. But Shimomura had no idea what type of molecule the luciferin was. Was it a protein, a sugar, a nucleic acid, an amino acid, or possibly a previously unknown structure?

Shimomura faced a monumental task, even by today's standards. The job required him to isolate and purify only the luciferin molecules from the mixture of tens of thousands of different molecules that make up *Cypridina*. To make matters worse, the luciferin is extremely unstable and quickly degrades when exposed to oxygen. Harvey's group had developed a partial purification method that resulted in a highly concentrated *Cypridina* luciferin, but their solution was not pure enough to obtain crystals. If impure preparations were used to study the physical and chemical properties of a reaction, it would produce false results that could send Shimomura down a research dead-end.

In the spring of 1955, Shimomura set out to obtain crystalline luciferin from *Cypridina*. He started with the existing protocol pioneered by Harvey's group and made several refinements that resulted in an extract with higher yield and greater purity. He found that the more religiously he excluded oxygen from his preparations the greater his yield. He took this approach to an extreme by performing all of the purification steps in a hydrogen environment. Given the explosive nature of hydrogen gas, such experiments were extremely dangerous. Each of the purification attempts required seven days and nights of continuous work. Heedless of the peril, Shimomura persevered, but after 10 months of producing preparations of greater and greater purity he had still failed to produce a luciferin crystal.[8]

Then one day, frustrated by yet another failed experiment, Shimomura left a small amount of purified luciferin out on the bench overnight in a strongly acidic medium. When he looked at the preparation the next morning, to his amazement he saw that small red crystals had formed in the solution. His serendipitous oversight had produced crystals of pure luciferin. Treatment with acid was the magic trick that allowed the crystals to form. The luciferin activity of the crystals was 37,000 times as intense as the dried animal per weight, indicating it was 20 times more pure than the material produced by the Princeton group. For the 27-year-old Shimomura, this was an enormous achievement. "Even if my success was accidental, it gave me self-confidence, and a feeling of 'if it is not impossible, then I can do it.'"[9]

Shimomura went on to characterize the crystalline luciferin, demonstrating many of its basic chemical properties. It would take almost 10 years, however, before the exact nature of *Cypridina* luciferin was established. Eventually Shimomura and his collaborators would determine the chemical structure of *Cypridina* luciferin. In 1957 Shimomura published his first major paper in the *Bulletin of the Chemical Society of Japan*, entitled "Crystalline *Cypridina* Luciferin."[10] Aware of the difficulties that had plagued other researchers for decades, Hirata was surprised at Shimomura's success.

Meanwhile at Princeton in 1957, one of Harvey's many students, Frank Johnson, began moving into the senior scientist's research area.

Osamu Shimomura (left) and a coworker, Toshio Goto, obtaining the first crystals of luciferin at Nagoya University in Japan. From the newspaper *Chunici* (Nagoya, Japan), circa March 1956.

"When Harvey retired, leaving a number of basic problems unsolved, I elected to seek some of the answers," he wrote. This included purifying *Cypridina* luciferin. That year, unaware of Shimomura's progress, he traveled to Japan's Izu peninsula to work with live specimens of *Cypridina* "to avoid the multitude of impurities inevitably present in extracts of whole, dried specimens." Johnson wrote that although "it should have been possible to obtain pure luciferin . . . technical difficulties were encountered in the lack of specialized equipment and the high temperature of the laboratory during the summer months."[11] Although unable to purify the luciferin, he did learn of Shimomura's paper upon returning to the United States. Frustrated by his own attempts to purify the molecule, Johnson was impressed by the young chemist's achievement and invited Shimomura to come and work in the United States. Shimomura accepted.

Though Shimomura was not a student at Nagoya University, Hirata offered him a Ph.D. as a going-away present when he heard of Johnson's offer. Hirata was familiar with the American university system, having worked a year in Louis F. Fieser's laboratory at Harvard University in 1952. He knew that a doctorate would double Shimomura's starting salary from $300 to $600 a month. After completing the necessary forms and paying only $10 in processing fees, Shimomura was now a newly minted Ph.D., having got it solely on the basis of his accomplishment in crystallizing *Cypridina* luciferin. "The Ph.D. was not in my path. I was just trying to finish my job," Shimomura said. In an ironic twist, it was the massive wartime collection of *Cypridina* that gave Shimomura the quantities of the animal that he needed to work out the purification procedure. Having only small quantities of the imported dried animal to work with had always hampered the Princeton group. This byproduct of a war that nearly devastated Shimomura's life launched his scientific career.

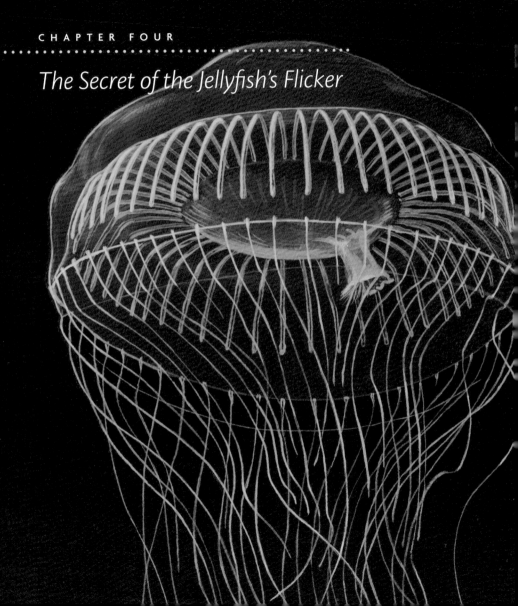

CHAPTER FOUR

The Secret of the Jellyfish's Flicker

BY THE TIME Osamu Shimomura received his invitation to Princeton in 1959, Edmund Harvey's health was deteriorating and Frank Johnson had begun to take over operations at the Princeton bioluminescence lab. Harvey died on July 21, 1959, just 14 months before Shimomura's arrival. Although Johnson had offered to pay for Shimomura's travel expenses, the ambitious scientist had applied for and received a Fulbright travel grant that included transport to the United States along with several weeks of English lessons. In August 1960 Shimomura left Yokohama bound for the United States aboard the M.S. *Hikawa Maru*, a 12,000-ton luxury ocean liner known as the Queen of the Pacific, the only large Japanese ocean liner to escape destruction during World War II. Shimomura recalls: "The ship left the pier with thousands of colored tapes draped between the ship and people on the pier. It was an unforgettable experience. The ship took 13 days to cross the Pacific, south of the Aleutians, finally arriving in Seattle. Then it took three more nights to cross the continent by rail, on a Pullman car. I remember this trip as my first trip outside Japan and the most luxurious trip in my life."[1]

When the train pulled into the semirural town of Princeton Junction,

Japanese Fulbright scholars of 1960, including Osamu Shimomura (back row, eighth from the right). Courtesy of Osamu Shimomura.

New Jersey, almost three weeks after Shimomura had left Japan, Johnson greeted Shimomura on the platform and took him to the laboratory in Guyot Hall—a Tudor Gothic–style building rimmed with gargoyles. There, Shimomura found himself in a familiar situation. Johnson led him into a darkroom and handed him a jar of dried white powder of the bioluminescent jellyfish *Aequorea*.[2] Johnson mixed the powder with wa-

Opposite: The jellyfish *Aequorea victoria*. Photo by Claudia Mills.

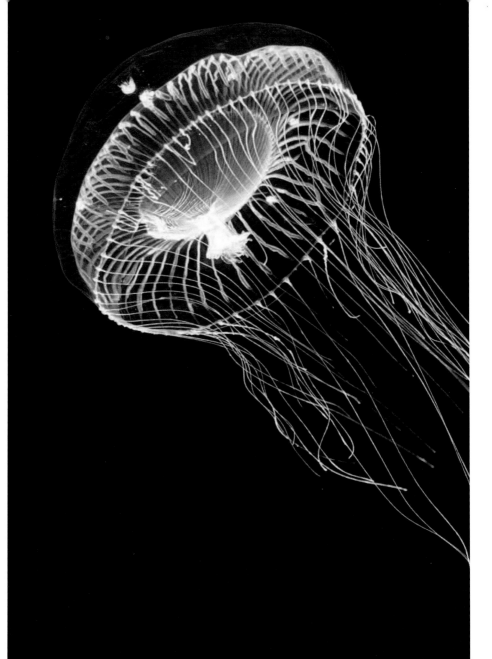

ter, expecting it to glow, but contrary to Shimomura's earlier experience with *Cypridina*, the room remained dark. Johnson followed this failed demonstration by telling Shimomura of the jellyfish's abundance off the island of Friday Harbor, Washington, in Puget Sound. Inside the darkroom, Johnson asked: "Would you be interested in studying this jellyfish?"[3]

"Yes, I will be happy to study the jellyfish," replied Shimomura, having no preference for what he would work on next.[4]

Aequorea grows to a diameter of three to four inches and is shaped like an umbrella with approximately 100 bright green glowing pinhead-sized specks spaced around its outer rim. Of all the bioluminescent animals that Harvey and Johnson studied, this was one for which the Dubois luciferin/luciferase experiment repeatedly failed. If a jellyfish is lysed, a process similar to being spun in a blender, the puree continues to glow for a few hours before fading. Once the bioluminescence was exhausted, it could not be reactivated under any circumstances.

Several months later, in June 1961, Johnson and Shimomura—along with Akemi Shimomura and Johnson's assistant, Yo Saiga—piled into Johnson's station wagon and headed west, through Chicago and Glacier National Park to the Friday Harbor Laboratories. Johnson, the only licensed driver, did all the driving, putting in 12-hour shifts throughout the seven-day journey. When they reached Anacortes, a seaport 70 miles north of Seattle, they took a two-hour ferry ride to San Juan Island, part of an archipelago that lies between the mainland and Vancouver Island. During the ferry ride to the laboratories, Shimomura looked over the side and was amazed to observe at first hand the legendary jellyfish's abundance. A constant stream of voluptuous orbs floated gently alongside the ferry's hull.

They were greeted on the island by Robert Fernald, then the director of the laboratory. Established in 1904, the laboratories consisted of only a few wooden cabins tucked into the sloping hills of the remote island. Fernald brought his visitors directly to Lab 1, a small building consisting of two rooms, which they were to share with three other researchers and Ghillies, a Scottish deerhound. The dog was the "lab assistant" of the outspoken University of Washington zoology professor Dixy Lee Ray. Ten years later Ray would be named chair of the Atomic Energy Commission by President Richard Nixon, in 1976 she would be elected the first woman governor of Washington state.[5] In 1961, however, her focus was the biology of wood-boring worms.

Johnson and Shimomura immediately unpacked their equipment from the station wagon, including a bulky two-foot cubed integrating photometer, set up the laboratory, and got to work. Using shallow nets similar to those used to clean swimming pools, they began to scoop up jellyfish at the pier in front of the laboratory. They filled bucket after bucket, one jellyfish at a time. Using a method developed by Harvey 40 years earlier, when he had first encountered the jellyfish, they removed only the light-producing organs for analysis.[6] The light organs, or photocytes, are distributed evenly along the brim of the umbrella. This arrangement allowed them to easily snip off the brim with a pair of scissors, creating a thin ring containing concentrations of light organs. Johnson and Shimomura then squeezed these rings through a cotton handkerchief, yielding a liquid they creatively dubbed "squeezate." This viscous jelly gave off light for several hours before the light-producing cells were cytolyzed and the reaction was exhausted.[7] Over the summer of 1961 they would collect and dissect over 9,000 jellyfish.

With the squeezates in hand, they began the task of isolating the luminous reactants. Their approach to purifying the bioluminescent components of *Aequorea* was similar to the one Shimomura had used to ob-

Collecting the jellyfish *Aequorea victoria* off the dock of Friday Harbor Laboratories, Washington, summer 1974. Left to right: Akemi Shimomura, Joseph Chang, Osamu Shimomura, Mrs. Chang, Mary Johnson, Frank Johnson, an unidentified collection assistant, Tsutomu and Sachi Shimomura. The orange and black tape on the poles of the jellyfish nets display Princeton's colors. Courtesy of Osamu Shimomura.

tain the *Cypridina* luciferin. This approach assumed a two-component reaction, and consisted of halting the reaction and then separating the luciferin and luciferase as soon as possible. Tactics to halt the reaction included removing, inactivating, or consuming one of the necessary components. Once one component was suppressed, Johnson and Shimomura thought, the remaining component could be purified.

They tried anything available on the island: various salts, metals, proteins, enzymes, and even laundry detergents. *Aequorea,* however, defied all of Johnson and Shimomura's attempts. After several days of collection and experimentation, they were unable to separate the reactants—and this was only the first step toward isolating the luciferase and determining the mechanisms of jellyfish bioluminescence. Out of ideas, Shimomura abandoned the laboratory and went into a deep meditative phase. He took advantage of the laboratory's peaceful summer location surrounded by stands of Douglas fir. For several days, Shimomura left his experiments behind and focused all his attention on what could be causing the jellyfish to glow. "If there was no luciferase, what could possibly be driving the reaction?" he wondered. A small rowboat served as his place for meditation. Day after day, he rowed into the calm harbor, his bare feet dangling over the boat's edge as he lay supine under the warm summer sun. At times, he would fall asleep and be swept away by the tidal current. One afternoon while floating on the harbor waters, he awoke with an idea that jolted him in its simplicity. Shimomura summarized the thought as follows: "Even if a luciferin-luciferase system is not involved in the jellyfish luminescence, another enzyme or protein is very likely involved directly in the light-emitting reaction. If so, the activity can probably be altered by a pH change, at least to some extent. Indeed, there might be a certain level of acidity at which an enzyme or protein could be reversibly inactivated."[8]

Quickly he rowed back to the laboratory, eager to begin a new series of experiments to test his hypothesis. Upon his return, however, Shimomura was disappointed to find that Johnson rejected the new model and refused to abandon the search for a classic luciferin-luciferase type reaction. A fractious and awkward relationship began to develop between the two scientists as Shimomura set about testing his new idea.

Working alone in a small corner of the building, he pursued his theory that something else was responsible for the jellyfish's glow. Meanwhile, Johnson and Yo Saiga continued, unsuccessfully, trying to extract a luciferin. As each day passed, the relationship between Johnson and Shimomura became more and more strained. Shimomura recalled: "I thought I would have to go back to Japan at the end of the summer."[9] Soon after, Johnson abandoned his work. Frustrated that he could not find a solution, he retreated from the laboratory and could be found sipping cocktails with Dixy Lee Ray.

Shimomura began making fresh batches of squeezate. He mixed this jelly with various weak bases and acids, but the bioluminescence persisted. Shimomura then treated the squeezate with a solution buffered at pH 4, at which point it stopped glowing. He then slowly added baking soda (sodium bicarbonate) to bring the pH back to 7, neutrality. To his surprise the squeezate began to glow faintly as pH was raised. This meant it was possible to inactivate the light-emitting reaction, an important step in the purification process. Shimomura was excited by this first glimmer of hope, and was convinced he had overcome a major obstacle. In the process of cleaning up the glassware after the experiment he poured the neutralized squeezate solution into the laboratory sink. When the solution hit the sink it gave off an "explosively strong" flash of blue light.[10] Something in the sink had strongly activated the squeezate. It did not take long for Shimomura to determine that it was seawater in the sink that caused the explosive release of light. Knowing the composition of seawater, Shimomura soon identified calcium as the activator. Calcium is the fourth most abundant ion in seawater.

If calcium activated the reaction, then removing it could theoretically inhibit the reaction. Shimomura realized that if he could exclude calcium from the squeezate just after it was made, he could add it back later during the purification process to activate the reaction. With this new plan in mind, he devised an improved procedure using a calcium-sequestering substance called EDTA (ethylene diaminetetracetic acid) that sopped up all the calcium ions and inhibited the reaction five times more efficiently than acid. Shimomura ran to retrieve Johnson from his cabin. He demonstrated the calcium activation to Johnson, who gasped, "Oh my goodness!"[11]

Over the next year, Shimomura and Johnson used this technique to extract, purify, and characterize the properties of the jellyfish's unique bioluminescence system. The reaction did not appear to need a classic luciferin in order to occur; it merely needed seawater. The jellyfish's bioluminescent system consisted of a single component instead of the expected two. Using the EDTA protocol to inhibit the bioluminescence reaction, Shimomura purified a photoprotein he called Aequorin that when activated by calcium emitted bright blue light. He was the first person to describe this novel form of calcium-activated bioluminescence that required neither luciferin nor molecular oxygen, as all other known bioluminescent reactions did. Many years later it would be determined that, in fact, the reaction does involve a luciferin, a small molecule, called coelenterazine, that binds to the Aequorin rapidly and tightly but produces light only when calcium is present. This Aequorin/luciferin complex remains stable in the jellyfish's cells, where the concentration of calcium is very low. When calcium levels are elevated in the cell, the molecular complex completes the reaction and the jellyfish gives off light. The jellyfish uses intracellular calcium concentrations to control light production, enabling it to produce its characteristic flicker when disturbed. Once the reaction has taken place, the protein needs to be "recharged" with fresh luciferin in order to produce light again. Like all other bioluminescent reactions, this recharging process requires oxygen.

The calcium dependence of Aequorin's light emission was an intriguing new system. Researchers would later determine that Aequorin and other calcium-dependent photoproteins from relatives of *Aequorea* have very similar molecular structures. In essence, the protein part clamps around the bound luciferin when no calcium is present. When calcium ions arrive, they bind to two and possibly three different binding sites on the protein. At that point, the protein almost instantaneously undergoes a very subtle change in its shape. This structural change activates the Aequorin that causes the bound luciferin to undergo an internal chemical reaction that gives off light. The protein is like a gun with the hammer cocked; the calcium binding pulls the trigger. As chemical reactions go, ones that give off light as a major byproduct are highly energetic, which suggests that the energy difference between the bound luciferin and the luciferin product that has given off light is substantial. It remains a mystery how the Aequorin protein manages to hold the luciferin in the highly energetic state while waiting for calcium. This highly energized state ensures that the reaction—when it does happen—occurs very quickly in the form of a single flash of light rather than a slowly developing pulse. This flash helps to distinguish the animal from the background. Fireflies, which also need to achieve a rapid flash of light, use adenosine triphosphate (ATP)—the energy currency of cells—rather than calcium as a trigger in an otherwise similar reaction.

Although Shimomura and Johnson's identification of Aequorin arose from their study of the natural process of bioluminescence, it did not escape their attention that this protein was an exquisite sensor of calcium levels. Calcium is highly abundant in living organisms; most of it is sequestered in the cells, but a minute fraction drifts around freely, triggering important cellular processes. In a 1963 research paper, Shimomura proposed that Aequorin could be used as a calcium detector in a biological system, revealing the quantity of free-floating calcium by the amount of light given off.[12] Several years later, two researchers from the University of Oregon showed that Shimomura's idea was correct. Ellis Ridgway and Christopher Ashley were studying excitation-contraction coupling in muscle cells.[13] At that time it was understood that an electrical pulse in muscle fibers caused contraction of the fiber, and that this process required free calcium to enter the muscle cell. What was not clear was the temporal relationship between calcium increases in the cytoplasm of the muscle fiber and the contraction of the fiber.

Ridgway and Ashley traveled to Friday Harbor in 1967 and used Shimomura's method to purify Aequorin from several thousand jellyfish. They then injected the material into the colossal individual muscle fibers of the acorn barnacle *(Balanus nubilus)*, also found in Puget Sound. Next they electrically stimulated the injected cell while recording light emissions from the fiber. During each contraction they recorded a flash of light from the fiber. In between the pulses the fiber was dark. These pioneering experiments provided the first concrete evidence that calcium levels in muscle fibers are very low at rest and then rise and fall rapidly during electrical stimulation. Their efforts offered the first direct measurement of an ion-concentration flux in a living cell in real time. The rapid onset of the calcium pulse followed by a slightly slower decay indicated that the cell has elaborate methods of regulating its intracellular calcium levels. The current understanding of calcium dynamics in cells, and the tight temporal and spatial control that cells maintain over intracellular messengers, grew directly from these experiments. It is now known that all types of cells modulate their calcium levels, and that calcium dynamics control such important cellular processes as neurotransmission in brain cells, contraction of the heart, cellular division, and the release of insulin into the bloodstream.

Ridgway and Ashley's experiments were some of the first to borrow a compound from nature to reveal another unknown biological process. Millions of years of evolution have perfected a wide range of chemical and biological processes that cannot currently be designed from scratch by scientists. But existing compounds can be plucked from their original organisms and transferred to other systems as study tools. This type of borrowing from nature to study nature would become the foundation and lifeblood of the molecular revolution that would envelop science several years after Ridgway and Ashley's study with the barnacle.

What perplexed Shimomura from that first experiment was that the light flash he saw in the sink was blue—yet jellyfish always emit green light. In all bioluminescence reactions studied up to that time, the color of light produced by the animal was the same as that seen when the purified reaction was performed in a test tube. Shimomura surmised that the animal produced blue light by Aequorin and then this light was simply absorbed by a substance in the animal's tissues and reemitted as green light. This observation was first recorded as a two-sentence footnote, tucked six pages into a scientific paper by Shimomura, Johnson, and Saiga on the bioluminescence system of the jellyfish: "A protein giving solutions that look slightly greenish in sunlight and only yellowish under tungsten light, and exhibiting a very bright, greenish fluorescence in the ultraviolet of a Mineralite, has also been isolated from squeezates. No indication of a luminescent reaction of this substance could be detected."[14] While bioluminescence is the production of light, fluorescence is the conversion of one color of light to another.

In a subsequent paper, Shimomura and Johnson presented this idea, and showed spectroscopic data to support the differing color emissions. They were also aware that the light organs of *Aequorea* emitted green

To Dr. Osamu Shimomura –
My good friend and respected colleague –
From Frank H. Johnson
Princeton, New Jersey, USA, 1976

Frank Johnson examining a glowing beaker of bioluminescent bacteria, Princeton University. Courtesy of Osamu Shimomura.

fluorescent light when exposed to ultraviolet light, as investigators at Friday Harbor Laboratories had shown seven years earlier.[15] Why the animal has developed a system to produce green light instead of blue remains a mystery.

What Shimomura discovered in Friday Harbor was not only the workings of light generation in the jellyfish *Aequorea,* but also a second protein: a green fluorescing protein (GPF). Miraculously, over the next 30 years, this obscure protein would be transformed into a tool widely used in biomedical research. Over 4 decades later, Shimomura would remark: "When GFP was discovered, the brightness and beauty of the fluorescence certainly inspired some yet unknown applications, but the applications like the tagging of a protein in a living system were beyond our imagination at the time, probably not in the sight of anybody."[16]

CHAPTER FIVE

The Light at the End of the Rainbow

WHY IS THE SKY BLUE? How do clouds form? Where do the colors of the rainbow come from? These deceptively simple questions have perplexed people for thousands of years. One person, Sir George Gabriel Stokes, provided the answers to all of them. In addition, he also discovered the phenomenon of fluorescence. Born in 1819 in the coastal town of Skreen in northwest Ireland, Stokes, the youngest of six children, had a modest childhood. His father, Gabriel, a rector in the local church, provided early home schooling. Stokes then learned arithmetic from a local parish clerk who was often impressed when "Master George had made out for himself new ways of doing sums, better than the book." In addition to displaying early signs of mathematical dexterity, as a youngster Stokes was passionate and prone to occasional violent fits of rage. In 1832, at the age of 13, he was sent to school in Dublin for a more formal education. Before departing, his brothers warned him about his ebullient behavior and told him not to give "long Irish answers" or his peers would mock him. From that moment on, Stokes fostered a habit of answering most questions with a simple "yes" or "no." His sister Elizabeth Mary described Stokes during his school days: "Between the ages of sixteen or seventeen he was keen in the study of butterflies and caterpillars. One day while . . . returning from a walk, he failed to respond to the salutation of some ladies of his acquaintance; when asked the reason of such odd behaviour he answered that he could not bow, as his hat was full of beetles."[1]

At the age of 18 Stokes moved to Cambridge, England, where he would remain almost without intermission until his death 66 years later. Even when he was an undergraduate at Cambridge's Pembroke College, his mathematical prowess was attracting the attention of leaders in the field. He won the prestigious Senior Wrangler award, given to the top student in each class, and immediately following graduation, in 1841, he was appointed a Fellow at the College.

The groundwork for modern physics was laid during the mid-1800s. Physical scientists of that era created the disciplines of electricity, acoustics, optics, hydrodynamics, and chemistry. Hypothesis-driven discovery had become standard, and scientists used logic and math to describe the physical world. Lord Rayleigh (1842—1919), who developed laws of optics and light scattering, and Michael Faraday (1791–1867), who developed laws of electricity and magnetism, were two of Stokes's contemporaries. Lord Kelvin was another contemporary, colleague, and fellow Irish mathematical prodigy. Admitted to Glasgow University in 1834 at the age of 10, he invented refrigeration, devised the scale of absolute temperature that still bears his name, invented devices that allowed telegrams to be sent across oceans, and postulated the Second Law of Thermodynamics. These scientists used the type of logical analysis that Sir Isaac Newton had used to co-invent calculus and describe light, motion, energy, and gravity.

Within three years of being elected a Fellow at Cambridge, at the age of 24, Stokes published his theory of the viscosity of fluids, "On Some

Portrait of Sir George Gabriel Stokes after receiving the Senior Wrangler award. Courtesy of Pembroke College, Cambridge University.

Cases of Fluid Motion," a work that redefined hydrodynamics.[2] This paper "contained a beautiful mathematical solution of the problem of finding the motion of an incompressible fluid . . . which constituted the complete foundation of the hydrokinetics," Lord Kelvin wrote.[3] Stokes later summarized this work in *On the Theories of the Internal Friction of Fluids in Motion*.[4] Several other articles followed, such as "On the Effect of the Internal Friction of Fluids on the Motion of Pendulums" in 1850.[5] This paper, whose basic conclusion became known as Stokes's Law of Settling, described the rate at which particles move through fluid media. In addition, Stokes determined the oscillations of a viscous fluid moving uniformly at small velocities. In doing so, he explained how clouds remain in suspension, by determining the terminal velocity of water vapor falling through the air.

In 1857, at the age of 38, Stokes married Elizabeth Haughton and lost his fellowship at Pembroke College owing to a stipulation against marriage. (His fellowship was later reinstated when the prohibition on marriage was revoked.) During their engagement, Stokes wrote sprawling letters, some over 50 pages long, to his fiancée. In one, he says that he was up until 3:00 A.M. wrestling with a mathematical problem and fears that she will not permit such behavior after their marriage. The two took rooms in Lensfield Cottage, situated in a large garden opposite the south side of Downing College, Cambridge. Stokes set up a makeshift optics laboratory at his home, and it was there that he made many of his greatest discoveries.

> A hole was cut in the window-shutter of a darkened room, and through this the light of the clouds and external objects entered in all directions. The diameter of the hole was four inches, and it might perhaps have been still larger with advantage. A small shelf, blackened on the top, which could be screwed on to the shutter immediately underneath the hole, served to support the objects to be examined, as well as the first absorbing medium. This, with a few coloured glasses, forms all the apparatus which it is absolutely necessary to employ, though for the sake of some experiments it is well to be provided also with a small tablet of white porcelain, and an ordinary prism, and likewise with one or two vessels for holding fluids.[6]

In "Dynamical Theory of Diffraction," Stokes provided a full mathematical explanation of the propagation of motion in a homogeneous elastic medium.[7] He concluded that the plane of polarization is the plane perpendicular to the direction of light's vibrations. As light moves through the atmosphere, most of the longer wavelengths, such as red, orange, and yellow, pass unimpeded. But shorter-wavelength light, such as blue, is absorbed by the gas molecules in the atmosphere and radiated in different directions. When we peer up at the sky, our eyes perceive the scattered blue light. "We now consider it one of the surest truths of physical science," Lord Kelvin later wrote of Stokes's extraordinary breakthrough.[8]

Working from his makeshift laboratory, Stokes went about studying the properties of sunlight. He found that when he placed a glass prism in a beam of sunlight coming through the window-shutter, it produced a rainbow of colors on the wall. This rainbow, originally discovered by Newton, consisted of bands of color always in the same order: red, orange, yellow, green, blue, and violet. When Stokes placed pieces of stained glass from a nearby church in front of the prism, the resulting spectrum contained only the color of the glass. White light from the sun actually contains all the colors of the rainbow, but the stained glass absorbs most colors, allowing through only light that is the color of the glass. On April 28, 1852, Stokes placed pieces of blue glass over the hole in the window-shutter, producing blue light. Next he placed a glass beaker containing a yellow-colored quinone solution into the blue light beam. Quinone is a fluorescent chemical found in plants. When illuminated by the blue light, it produced a yellow glow. Stokes systematically examined a range of chemicals and solutions and found that while many absorbed or filtered light, only a few produced a color of light different from the color of the illuminating beam. He called this phenomenon fluorescence.

Materials that exhibit fluorescence absorb light of one color and emit it as a different color. In all cases the new fluorescent light produced was shifted in color toward the red end of the rainbow compared to the light used to illuminate the specimen. Absorption of blue light produced green light, while absorption of green light produced red light. Stokes used these findings to formulate what is now known as Stokes's Shift, which holds that the emitted fluorescent light has less energy (is red-shifted) than the exciting light. In a paper entitled "A Discovery" Stokes wrote: "A phenomenon of internal dispersion, a ray of light actually changes its refrangibility."[9] "Refrangibility" is an antiquated term used to describe the colored nature of light. Stokes summarized his findings in a communication to the Royal Society of London on May 27, 1852, entitled "On the Change of the Refrangibility of Light."[10]

Though he understood little of the nature of light, Stokes described the basic physical properties of fluorescence. Not until almost 75 years after Stokes's discovery of fluorescence was a complete description of light formulated. It had been understood since the 1600s that light was a form of energy, but many of its properties seemed to defy a unifying description. It seemed to be composed of "tiny particles, or corpuscles," and that is how Newton described it in the *Opticks* in 1704.[11] Light always arrives in discrete packets, today called photons. These packets seemed to emanate outward from a light source in a random spray. Molecules absorb light energy in discrete packages, each packet delivering a relatively consistent amount of energy to the molecule at a given wavelength. But light often also behaves like a wave. Light waves can interfere with and cancel one another; light can be polarized; and it has oscillatory properties best described by wave functions.

For over a century there was a vigorous debate over whether light consisted of particles or waves. Resolution came with the development of quantum mechanics. Quantum mechanics not only encompassed all the observed properties of light but also allowed the dynamic behavior of subatomic particles such as electrons, neutrons, and protons to be studied and quantified. Like relativity, quantum mechanics offers a new way of looking at the world. It explains phenomena not as definite events and particles but as probabilities of occurrence. Erwin Schrödinger, an Austrian physicist, used a simple experiment to describe the theory. He postulated a living cat in a chamber along with a capsule of deadly poison. The chamber also contains a tiny amount of a radioactive substance. If even a single atom of the substance decays while the cat is in the chamber, the vial will crack, releasing the poison and killing the cat. While the experiment is taking place, it is impossible to know whether or not the atom decayed. Observation or measurement would interfere with the experiment. According to quantum law, the cat is both dead and alive—in a superposition of states. This situation is sometimes called quantum indeterminacy or the observer's paradox.

Quantum mechanics requires an entire book for a complete explanation, but to understand the processes of fluorescence and bioluminescence, only a few important properties of light need to be understood. A tungsten light bulb, for example, consists of an electrically conductive filament, usually a thin metal wire, suspended in a vacuum. When voltage is applied, electrons—the electrical current—flow through the wire. As the voltage increases, so does the stream of electrons. When electrons pass through the wire they bump into the atoms that make up the wire. As the electrons collide, each releases some of its kinetic energy as heat. That heat then agitates the atoms in the wire. Agitated atoms give off their energy in the form of both light and heat. Once agitated, an atom will release a photon of light. The atoms constituting the wire receive different amounts of energy, depending on the force of the passing electrons. The energy of the photon of light is described as its wavelength and is determined by the energy contained in the agitated wire atom. Short wavelengths of light vibrate faster and have higher energy. Long wavelengths of light vibrate more slowly and have lower energy. The intensity of light depends on the number of photons emitted by the wire over a given period of time. Brighter light means that more photons are generated and reach the observer.

The physical properties of light are loosely analogous to those of sound. The brightness of light is like the volume of a sound while the wavelength of light, or color, is like pitch. Shorter wavelength sounds are more treble and longer wavelengths are more bass. In the same way that the ear perceives different sound waves as different pitches, the eye perceives light of different wavelengths as different colors. Color, therefore, is a psychophysical representation of wavelength and is not a physical property inherent to the light itself. The human eye can only see light from around 400 nanometers (a nanometer is a billionth of a meter) in wavelength—violet—to around 700 nanometers in wavelength—deep red. Light of longer wavelengths (700 to over 1,000 nanometers) is called infrared and light of shorter wavelengths (from 200 to 400 nanometers) is called ultraviolet light. The rainbow of colors that varies between 400 and 700 nanometers (violet, blue, green, yellow, orange, and red) forms the color spectrum.

The study of light energies and color, or spectroscopy, is used to define the number and range of different colored photons that are emitted by a particular light source. Thus the composition of an unknown object can be analyzed by virtue of how the object interacts with or produces light. Spectroscopy is a major tool astronomers use to analyze the composition, temperature, size, and even direction of movement of celestial bodies. Instruments called spectrophotometers take in light and, using optics

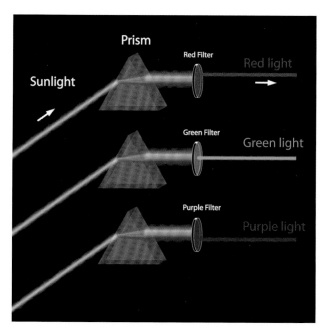

Sunlight is composed of all wavelengths of light. After passing through a prism the white light is broken apart into its component wavelengths, producing a rainbow of colors. A green filter blocks all other colors of light except green.

akin to a prism, break the light apart into its colored components. This light spectrum, visualized as a rainbow, is then analyzed and the energies of the individual colors quantified. The result is a spectrograph that describes the various energies of light.

The sunlight that entered Stokes's home laboratory contained photons of light with a range of energies from ultraviolet to infrared. This photon mixture appears as white light to the eyes. When he placed the blue stained glass in front of the sunlight beam, it absorbed all but the blue photons of light. Those blue photons passed through the glass and shined on his beaker of quinone solution. When light strikes molecules in an object or solution, it interacts with the molecules in several different ways. First, the molecules within an object or solution can simply absorb the photons. In most cases, the light energy is simply converted to molecular agitation, or heat, such as the warming rays of the sun. The surface of an object, such as a mirror, can also reflect light. In the case of mirrors, the wavelength and intensity of the light is largely maintained, although even the best mirrors absorb some energy. In other cases, light can largely pass through an object, as with glass. When photons pass through an object, they will be refracted, or bent, and their direction of propagation altered. This refraction is caused by changes in the speed of light as it passes through different substances. Lenses allow us to control refraction with great precision, correcting vision and letting us see deep into space. The light from very small objects can be enlarged by bending the light waves through lenses, as is the case with a magnifying glass, telescopes, and microscopes.

When white light containing photons with a wide range of energies reaches an object, the apparent color of the object depends on how the photons of light interact with it and which photons are returned to the viewer. The properties of the molecules constituting an object determine what types of photons the viewer sees. When white light is projected on an apple, the pigments in the skin of the apple absorb most of the light, except red photons. These photons are reflected back from the apple's skin and, therefore, the apple appears red. When white light is shown through red filter glass, by contrast, pigments in the glass absorb all of the different color light photons *except* the red photons, which pass through the filter.

Fluorescence is a special chemical property in which a molecule absorbs light but then reemits light back at a lower energy level, or color. At the molecular level, a fluorescent molecule absorbs a photon of light and absorbed energy causes an electron in the molecule to jump to a higher, more energetic orbit around the nucleus. That electron remains in this excited state for only a few billionths of a second and then falls back to the lower-energy orbit. Upon returning to its initial state, the electron emits a photon of light.

We are surrounded by fluorescent objects. All neon paints have fluorescent pigments. The glow of T-shirts and our teeth under black light is fluorescence. Black light is largely composed of ultraviolet light, invisible to the eye. However, many materials will absorb ultraviolet light and reemit it as lesser-energy light that falls within the range visible to the human eye.

Forensic technicians use ultraviolet fluorescence to find fingerprints at crime scenes. When human skin touches most surfaces it leaves behind trace amounts of amino acids. A chemical called ninhydrin (brand name Luminol) is applied to the surface of objects at crime scenes to detect these amino acids. When amino acids left by the perpetrator react with ninhydrin, they create fluorescent molecules that give off visible light when stimulated by ultraviolet light. The crime scene is then darkened and the sprayed surfaces are examined for glowing fingerprints with a hand-held ultraviolet lamp. This technique is much more sensitive and useful than the traditional black-powder method of detecting fingerprints.

Biologists quickly embraced the power of fluorescence to reveal the unseen. As scientists peer into smaller and smaller compartments of cells, traditional light microscopes become ineffective. Many interesting aspects of cells, their proteins and chemicals, are invisible to such microscopes. In contrast, even a single fluorescent molecule can be easily visualized because it produces photons of light with a unique signature. When that single molecule is exposed to intense excitation light, it will convert the absorbed photons into another wavelength that the viewer then sees. Light filters can screen out the excitation light, allowing a researcher to see the single fluorescent molecule as a pinpoint of light. Single molecule localization in cells is difficult because there can be billions to trillions of molecules per cell, but it is possible.

In the 1940s fluorescence caught the eye of Albert Hewett Coons, a histologist working as a medical resident in a hospital in Berlin, Germany. In his spare time, Coons began researching rheumatic fever. At that time, a great debate surrounding rheumatic fever concerned whether or not infected patients' lesions contained the *Streptococcus* bacterium. To prove whether the bacterium was present or absent, Coons developed an antibody directed against *Streptococcus* that had been tagged with a fluorescent label. Antibodies are proteins produced by white blood cells and are part of the body's immune response system. When foreign proteins enter the body, certain circulating white blood cells are activated to secrete large quantities of specific antibodies. Humans have thousands of different antibodies floating in the blood at any given time. Each different antibody has a "binding domain" that allows it to attach to only one specific bacterial protein. When the antibody encounters the bacterium, it attaches to the bacterium and inactivates it. Coons knew of the incredible specificity antibodies have for their target. Of the thousands of different proteins in the human body, antibodies against the bacterium of rheumatic fever will stick only to that bacterium. He envisioned using this specificity as a tool to locate the bacteria in tissue sections from patients with rheumatic fever.

The plan was to inject a particular antigen, in Coons's case a heat-killed *Streptococcus* bacterium, into a rabbit. The rabbit's immune system would then produce large quantities of antibodies against the bacteria. After a few weeks, he drew a small amount of blood from the rabbit and purified the antibodies from the plasma. Coons initially proposed to chemically link a colored molecule to the antibodies and then incubate these antibodies with thin sections of *Streptococcus*-infected tissue. Under a microscope, those lesioned parts of the tissue that contained the *Streptococcus* bacteria would bind the colored antibody and Coons could determine the location of the bacteria by visualizing the accumulation of colored antibodies.

The problem with this first approach was that not enough antibodies gathered at the lesion site to produce visible color. Coons had another idea. He decided to couple a highly fluorescent molecule called fluorescein to the antibody. This molecule glows bright apple green when illuminated with blue light. When the fluorescien-labeled antibodies were incubated with the bacteria-infected tissue, the antibodies accumulated only at those sites in the tissue where the bacteria were present. Then Coons used a special microscope to visualize the location of the bound fluorescent antibodies. For fluorescein-tagged antibodies, he illuminated labeled tissue sections with a bright blue light and then viewed the sections with a special filter that blocks blue light from reaching the viewer's eyes. In the absence of fluorescent antibodies, the tissue looks black because the blue light illuminating the tissue is blocked from view. But if the blue light strikes a fluorescein molecule, it is absorbed and converted to green light that readily passes through the green filter to the viewer's eyes. Consequently, regions on the tissue section that contain the bacteria, and therefore the fluorescent antibodies, glow a bright green. Coons co-opted the high specificity and binding affinity of antibodies as a way to visualize the location of a protein in a cell. The technique was very successful, and Coons managed to settle the argument—the rheumatic fever lesions were packed with the *Streptococcus* bacterium.[12]

Coons's technique, called immunocytochemistry or immunofluorescence, became an instant hit. Droves of scientists since then have used the method to localize all forms of proteins in cells. Additional fluorescent labels were developed in different colors to allow multiple labeling. AMCA is a blue fluorescent label; rhodamine, Texas Red, and Cy3 are red fluorescent tags; and Cy5 is a far-red fluorescent tag, to name a few. An enormous business has grown up around the production of specific antibodies against cellular proteins. One commercial supply house, Accurate, Inc., boasts 33,000 different antibodies. Immunofluorescence is the premier way to localize proteins and other cellular components at the microscopic level. And with multiple fluorescent dyes, a researcher can localize up to four different antigens in a single cell.

The popularity of immunofluorescence drove the production of high-grade light microscopes capable of fluorescent imaging. In the 1970s and 1980s, the major microscope manufacturers made significant advances in fluorescent microscope design. Three major microscope innovations would dramatically increase the quality of fluorescent images. First, Johan Ploem developed the epifluorescent microscope. Ploem used the same light path through the microscope to both excite and view the fluorescence of a specimen. This innovation allows a tissue section to be illuminated and viewed from the same direction, which significantly enhances the brightness and quality of the image.

The next major fluorescent microscope improvement was the laser scanning confocal microscope, invented by an unlikely character. Marvin Minsky, a pioneering cognitive neuroscientist, patented the design in 1961, but it did not enter commercial production until the late 1980s.

The microscope was a radical break from traditional epifluorescence microscopes. Standard fluorescent microscopes are limited by the blurry background generated by the out-of-focus regions above and below the focal plane in an image. A confocal microscope uses a laser to illuminate only a very small spot in the field of view. This single point of laser light is then rapidly scanned across and down a tissue section. The fluorescent light originating from the point source is collected and used to form a complete image in a computer. The microscope uses traditional excitation and emission filters. Using a point source allows both the excitation light and the fluorescent emission light to pass through a pinhole before reaching the viewer. Through a trick of optics, the passage of the light through the pinholes removes most of the out-of-focus fluorescent light from above and below the focal plane. This microscope produces stacks of thin image slices that can then be compiled to produce a three-dimensional picture of the tissue with exceptional clarity.

A more recent advance in fluorescent microscopic imaging is the multiphoton fluorescent microscope, the type of microscope used by the scientist in the Prologue to explore Alzheimer's plaques in the living mouse brain. Multiphoton microscopes form an image in a way radically different from that used by all other fluorescent microscopes. The confocal microscope excites out-of-focus fluorescent structures in tissue but does not allow this fluorescence to reach the viewer. A multiphoton microscope achieves the same effect but through a different approach. Like a laser scanning confocal microscope, a multiphoton microscope generates its image by scanning a laser beam across the microscope's tiny field of view. But the multiphoton microscope excites only a very thin plane in the tissue. This limited excitation is achieved by the use of a special laser, called a femtosecond-pulsed pumped-laser. This type of laser delivers a stream of very short light pulses, each flash lasting a thousand trillionth of a second. To put this into perspective, a thousand trillionth of 4.6 billion years (the time since the earth formed) is under 2.5 seconds. Although each pulse has very high energy, the laser does not damage cells because the pulses are separated by relatively long periods of darkness. Yet, at the point in the tissue where the laser is in focus, the photon flux density (how many photons are striking the tissue) is very high during the short light pulses—so high, in fact, that if the density was not pulsed the laser would instantly vaporize the tissue (and a part of the microscope). The intense energy levels alter the normal fluorescence process because a second photon collides with the molecule before it has had time to reemit the excess energy as light. The excited electron, which has now been raised to an even higher energy level, actually emits fluorescent light at a higher wavelength than the excitation light. This special technique allows the microscope to produce clearly defined images of structures deep in seemingly opaque tissues.

Immunofluorescence, combined with these modern microscopes, provides a way of localizing proteins to specific cells and even to specific locations within the cell. But because antibodies are large proteins that cannot penetrate the intact cell membrane, immunofluorescence is carried out only on nonliving cells.

The development of fluorescent histological techniques allowed for significant advances in the understanding of the compartmentalization of proteins within cells. The antibody techniques illuminated the highly organized cellular communities and the proteins that define them. But even though such histological techniques produced substantial achievements, scientists quickly found the use of dead tissue or cultured cells limiting. After all, the study of biology is the study of cell function in living organisms. Confocal and multiphoton microscopes provide the means to image fluorescent dyes or labels deep within living tissues without being invasive. But the antibody and reporter dyes cannot easily penetrate living tissue. In addition, these methods required the addition of foreign material—antibody, dye, and so on—that would have to pierce the tissue and cells to reach the intracellular targets. Inevitably, antibody dyes and reporter cells were destructive to the cells of interest, and therefore a less-than-ideal way to study life.

CHAPTER SIX

Illuminating the Cell

OSAMU SHIMOMURA'S 1961 discovery of green fluorescent protein took place at the beginning of a biological revolution in the field of molecular biology. Before that time, in order to study protein function it was necessary to purify the protein and then study its properties in a test tube. That is how Shimomura examined Aequorin. But since cell function requires collaborative interactions among thousands of different proteins, trying to understand cell function by studying an isolated protein is a bit like trying to deduce how an engine works by examining a single spark plug. You might learn something about the spark plug, but you probably wouldn't figure out how a car operates. It would be more worthwhile to examine an intact running engine. The molecular revolution gave scientists the crucial ability to study proteins in living cells.

One of the most significant early milestones of the molecular biology revolution took place in a small brick-lined room in the Cavendish Laboratories at Cambridge University in 1953. There, a 36-year-old graduate student, Francis Crick, and a 24-year-old postdoctoral fellow, James Watson, neither formally trained as a biologist, solved one of the great riddles of life. They did so by using X-ray diffraction data obtained from Rosalind Franklin and Maurice Wilkins, scientists working independently at King's College, London. In their office, Crick and Watson built a large model of DNA (deoxyribonucleic acid). On April 25, 1953, they published their work in a 900-word article in the journal *Nature* that described the molecular structure of DNA.[1] Seldom has the determination of a molecule's structure revealed such a succinct story. DNA is composed of a stretch of molecules linked end on end in an ordered sequence producing a molecular language. Suddenly a mechanism was revealed by which a cell could store voluminous information—blueprints—about how it is constructed. Previously, it was not understood how DNA functioned; delineation of its structure provided the first glimpse of how DNA was read by the cell to produce proteins. DNA's information is spelled out along the surface of two long and thin ropelike molecules that twist around each other, forming the famous double helix. The adenine (A) molecules on one strand are always paired to thymine (T) molecules on the other strand. Cytosines (C) are always paired to guanines (G).

DNA is a language where the four bases (A, T, C, and G) are the letters, and the genes, consisting of hundreds to thousands of base pairs in a row, are the sentences. Each gene contains the instructions necessary to make a specific protein. The entire complement of genes in an organism is known as the genome. Every cell in an organism contains a copy of the entire genome. The genome, therefore, can be thought of as a script in a theatrical play. Each cell has a copy of the entire play but every cell reads only its specific part, skipping over the lines that are not intended for it. From the moment of conception, for example, every one of the cells in your body has contained a complete set of your genes. Each cell has a copy of your entire genome, but a specific cell reads only lines or genes that define its character. Muscle cells read only certain portions of DNA, while brain cells read others.

The breathtaking discovery of the structure of DNA prompted microbiologists to formulate the central principles of molecular biology. Bit by bit, in laboratories throughout the world, scientists developed a core set of principles describing the biology of a cell. Within the cell nucleus, the DNA instruction set is read to produce short stretches of another nucleic acid, ribonucleic acid (RNA). These short strands of RNA are then decoded to produce proteins, the main machinery of the cell. Watson and Crick's description of the structure of DNA encouraged an army of scientists to begin deciphering how DNA is stored, read, and copied. Soon, it was realized that, with a few exceptions, the DNA → RNA → protein chain applied to virtually all forms of life, from plants to people, penguins to protozoa.

Two properties of DNA made it difficult to study. Each mammalian cell contains only two physical copies of each gene embedded in long stretches of DNA. Those genes are difficult to isolate, sequence, and study. Scientists needed a simple way to produce large quantities of DNA encoding for individual genes. The answer came in the form of *Escherichia coli,* a bacterium commonly found in the human gut. Bacteriologists had already developed techniques to isolate individual bacteria and expand them rapidly into pure cultures. A liter of culture could produce hundreds of billions of bacterial cells in only a few hours. In addition to having their own DNA, *E. coli* (and other bacteria) sometimes possess small separate circular pieces of DNA called plasmids. Plasmids represent a primitive form of sexual reproduction for otherwise asexual bacteria. Using tiny structures called conjugation tubes, the bacteria connect with one another to transfer plasmids. Scientist developed ways to slip a foreign gene into a plasmid, and then to reintroduce it into the bacteria, tricking the bacteria into producing a large amount of the foreign gene.

Another major advance was the discovery that bacteria express proteins called restriction endonucleases, or restriction enzymes. Each of these enzymes cuts a strand of DNA at a very specific sequence. For example, an enzyme called EcoRI cuts DNA only at the sequence GAATTC. These enzymes act as an immune system that snips DNA only from invading viruses, not from the bacteria's own genomes. Other enzymes were soon found that reproduce, cut, digest, and reconnect DNA strands. With the plasmids and enzymes at hand, scientists could cut up large strands of DNA into small, manageable pieces and insert them into plasmids so they could be amplified and studied. Then, using the bacteria's own machinery for assembling proteins, they could dupe the bacteria into producing massive amounts of protein from any gene.

The days of massive collections of animals for the purification of small amounts of proteins were over. Previously, purification involved pulverizing large quantities of a target organism. This was followed by lengthy processing procedure (sometimes taking days to complete), just to obtain a pinch of pure material. For example, Shimomura netted over 2.5 tons of jellyfish to produce between 100 to 200 milligrams of Aequorin.[2] With the new molecular approaches, the DNA sequence of almost any protein can be introduced into bacteria. These bacteria can then be stimulated to produce grams of almost any known protein overnight in a single laboratory. The animal where the gene originated never had to be collected again.

In addition, after determining a protein's sequence, scientists could now study how a particular protein functioned inside a cell. Until that point, protein function was largely a matter of biochemistry. It was studied by purifying the protein of interest, adding substrate, and watching

William McElroy extracting the luminescent compounds from fireflies, before the molecular biology revolution. McElroy was a leading bioluminescence researcher, a Ph.D. student of Edmund Newton Harvey, and director of the National Science Foundation from 1969 to 1972. Photo courtesy of J. Woodland Hastings.

the results. Now, protein function could be examined inside a living organism—rather than in a test tube. Newly cloned genes became widely disseminated and the knowledge of cell function leaped forward.

As scientists began to study multicellular organisms, they found that their genes contain massive amounts of genetic material not used to make protein. These portions, once thought of as "junk DNA," are scattered within the coding sequences so that 1,000 base pairs of

DNA coding for a protein may be spread over 10,000 total base pairs. The cell easily reassembles these fragments to produce the correct sequence. But it is very difficult for scientists to reconnect together the protein coding portions of the gene. Another biotechnology breakthrough solved the problem. While studying a type of leukemia caused by viruses, David Baltimore and Howard Temin independently discovered in 1971 that these pathogenic viruses possessed a unique property. Once the virus infected a mouse blood cell, it produced a copy of its RNA-based genome in DNA and then inserted this into the genome of the mouse. The mouse then unwittingly made copies of the virus DNA along with its own genetic material. This type of virus would become known as a retrovirus, similar to the AIDS-causing human immunodeficiency virus that was first described in 1983. This conversion of RNA into DNA was unique in the biological world and represented an inversion of the standard transcription of genetic information from DNA to RNA. Baltimore and Temin purified and characterized this unique enzyme, calling it reverse transcriptase.[3]

The biotechnological value of such an enzyme had not escaped Baltimore and Temin. And, sure enough, a powerful technique based on this enzyme was soon developed. Complemental DNA (cDNA) library production is a method whereby RNA from a cell is first purified in a test tube. In the process of making RNA, the cell edits all the nontranslated RNA, keeping only the protein's coding sequence. These short strands are purified and are then converted back into DNA through the use of reverse transcriptase. The scientists can then insert these RNA complementary DNA strands, or cDNA, into bacterial plasmids using a technique where each bacterium accepts only one gene. The collection of bacteria is then grown on a Jello-like substance made from seaweed, called agar. Thousands of colonies of bacteria begin to grow on the agar, each containing a cDNA copy from the cell of a single messenger RNA (mRNA), a molecule that acts as an intermediary between DNA and protein production. These colonies can be examined rapidly to determine if they contain a gene of interest. Once identified, a colony can be propagated to produce a pure culture of bacteria containing the desired gene. With this method, a single gene can be isolated from the thousands that are expressed in a given cell. The advantage of this approach is that cDNA clones contain only the information to make a particular protein, and not all the intervening sequences found in the genomic DNA. Temin and Baltimore received the Nobel Prize in Medicine or Physiology in 1975 for their discovery, sharing it with Renato Dulbecco, who pioneered the culturing of retroviruses.

By the mid 1980s the identification and characterization of new genes was commonplace. Laboratories throughout the world began employing molecular biology techniques to expand the study of biology.

These molecular techniques, however, were slow to be adopted by scientists studying bioluminescence and fluorescent proteins. As late as 1992 one of the only bioluminescence researchers applying molecular techniques to jellyfish was Douglas Prasher. After completing a doctorate in biochemistry at Ohio State University in 1979, Prasher accepted a position at the University of Georgia, studying bacterial genetics.[4] When his funding expired 4 years later, he moved to another group at the University of Georgia. The leader of the group, Milton Cormier, began studying bioluminescence in the 1950s with animals collected off the coast of Georgia. In particular, he was fascinated with an unusual and brilliantly bioluminescent animal, *Renilla,* or the sea pansy. Sea pansies are cnidarians and close relatives of jellyfish, anemones, and corals. The sea pansy possesses a single, giant polyp (about the length of a finger) that forms an anchoring stem—used to embed the animal in the sand. Its body has many anemone-like polyps, some for feeding and others to inflate the body. If the tide rushes out, leaving the animal exposed, it deflates and blends into the sand. The feeding polyps secrete a sticky mucous, used to ensnare small floating creatures. When disturbed, the sea pansy emits green bioluminescence from its polyps that permeates waves as they pass over the sea pansy's body. Cormier easily collected sea pansies during low tide on the beaches in front of the University of Georgia Marine Institute on Sapelo Island. Cormier's group also focused on a second favorite bioluminescent organism, the jellyfish *Aequorea*.

Since the characterization of the jellyfish's protein by Shimomura and Johnson, several laboratories, including Cormier's, had become interested in the study of the protein's properties and its use in research into cellular calcium dynamics. Many members of Cormier's group spent summers—alongside Shimomura—scooping jellyfish from the waters of Friday Harbor. They would then begin the arduous processing steps to obtain small amounts of purified material.

When Prasher arrived, Cormier recognized his molecular biology skills and asked him if he could isolate the gene for Aequorin. If Cormier had the gene, he could produce more Aequorin in his Georgia laboratory in one night than his team could purify over an entire summer at Friday Harbor. Once Cormier had the gene, his group would no longer have to collect thousands of the jellyfish for each experiment. Prasher's task of purifying the gene for Aequorin was difficult because each jellyfish had only scant amounts of Aequorin RNA, but in 1985, after spending 2 years collecting jellyfish and grinding them up for RNA, Prasher succeeded. He had cloned Aequorin and showed that bacteria could now be tricked into producing the bioluminescent protein.[5] "Those were the dark ages," Prasher later recalled. "It was a bear to sequence back then. Now you just buy a kit, check one time, and you're done. Quality control is already provided."

He was just in time. That year *Aequorea* mysteriously vanished from Friday Harbor and, as of the writing of this book in 2005, the jellyfish have not returned. "It is as if they realized they were no longer needed," says John Blinks, a researcher who had spent over a decade plucking jellyfish from the waters of Puget Sound.[6] In October 1987, Prasher accepted a position as assistant scientist at Woods Hole Oceanographic Institution (WHOI) in Massachusetts. But as a bacterial geneticist, Prasher found himself lost at an institute primarily interested in oceanography. "The biology department at the time wanted to bring in some molecular experience and I don't think I was the right person," Prasher recalled. Luckily, Prasher received a small grant from the American Cancer Society for his next project, to determine the sequence of green fluorescent protein.

Shimomura and Johnson, meanwhile, had shown little interest in studying the green fluorescing protein they had stumbled upon. For them, the point of purification was to isolate the chemical and physical properties of the bioluminescence reaction, not to focus on the fluorescent protein. Although they certainly thought it was odd that the jellyfish possessed a protein that lowered the energy of its natural light emission, their research program concentrated on other questions. For almost a decade after Aequorin's discovery, no one published a single article on the green fluorescent protein. Some researchers studying Aequorin even viewed the fluorescent protein as a "contaminant" because it mingled with the photoprotein in extremely high concentrations, altering the color from blue to green.

In 1971, while working on the physiology of another bioluminescent marine animal, *Obelia*, a colonial hydroid related to corals and anemones, two scientists at Harvard University encountered a green fluorescent protein. Woodland Hastings, a professor of biology and former student of Edmund Newton Harvey, and James Morin, a graduate student, were studying how *Obelia* produced its impressive nocturnal light show. Like a tiny delicate Christmas tree, *Obelia* attaches to the muddy bottom and creates synchronized light emissions from photocytes spread out along its body. It's not clear why the animal produces the elaborate light display, but Morin and Hastings were interested in how it produced the display. They found that the animal has a primitive nervous system that sends waves of electricity up its body and causes the photocytes to blink on and off. They were surprised at the notion of a fluorescent protein and investigated it further.

In two papers, they proposed that both *Obelia* and *Aequorea* transfer the light energy from the luciferins to the fluorescent protein by a special process called fluorescence resonance energy transfer (FRET), also called the Förster effect after its discoverer, the German physicist Theodor Förster.[7] Hastings and Morin proposed that Aequorin transferred its excited state energy to the fluorescent protein directly, without ever emitting a blue light photon. Although seemingly a trivial difference from Shimomura and Johnson's explanation, Hastings and Morin's account clarified why the jellyfish failed to emit blue light and revealed this novel feature of fluorescent proteins that would later prove very useful. Hastings and Morin also renamed it green fluorescent protein, or GFP.

Although Shimomura was not particularly interested in GFP, he did touch upon it in two publications in the 1970s: once to describe the crystallization of GFP in 1974 and again to elucidate the chemical structure of the fluorophore in 1979.[8] The first person to undertake the study of GFP as the sole focus of his investigations was William Ward, a biochemist and self-proclaimed workaholic. Ward's doctoral work concerned the purification and characterization of calcium-activated photoproteins from bioluminescent ctenophores.[9] These are fragile transparent, gelatinous, and nonstinging animals, also known as comb jellies. Ward was particularly interested in a genus called *Mnemiopsis*. This is a large and robust member of the comb jellies—about the size of an oblong orange—and does not fall apart when handled. After receiving his degree in 1974, Ward joined Milton Cormier's bioluminescence group at the University of Georgia. Cormier had a long-standing interest in the sea pansy *Renilla* and had several large coolers filled with animals no longer suitable for bioluminescence study. Storage destroyed the sensitive photoprotein but not the more durable green fluorescent protein. Ward's first paper, written with Cormier, describing the energy transfer between GFP and the photoprotein appeared in 1979, the same year Shimomura proposed a partially correct structure of the fluorophore.[10] That summer, Shimomura and Ward met at Friday Harbor. Shimomura says: "I learned in 1979 that W. W. Ward of Rutgers University, the pioneer of the isolation of the photosensitive ctenophore proteins, had been working on *Aequorea* GFP in addition to *Renilla* GFP. I thought my role was over and decided to discontinue my work on GFP. Dr. Ward established the complete structure of the chromophore several years later."[11]

Enamored with the beauty of green fluorescent protein, Ward selected it primarily on the basis of its aesthetic characteristics, and began characterizing the green fluorescent protein of both *Renilla* and *Aequorea*.[12] For nearly 15 years, he would remain the only person fully dedicated to the study of green fluorescent protein.

CHAPTER SEVEN

Glow Worms

COLD SPRING HARBOR LABORATORY, located 35 miles from Manhattan, on a secluded inlet off Long Island Sound, is normally a peaceful enclave for molecular biologists. Yet on one warm summer afternoon in 1954, an intensive bacterial genetics course had just concluded, and the students, from laboratories around the world, began their end-of-class festivities with a costume parade between Davenport Lab and Blackford Hall. Once they reached the dining area of Blackford Hall, the shenanigans began. Students dressed as viruses crowded together under a table draped with a white cloth meant to represent an infected and swollen bacterium ready to burst. The class was acting out how a lysogenic virus known as a bacteriophage attacks a bacterial cell. Viruses have no means of self-replication, so they hijack the bacterium's machinery, commandeering it to reproduce more viruses. Ultimately, the bacterium bursts, releasing multiple copies of the infecting virus. Sydney Brenner, a 27-year-old graduate student from Oxford's Physical Chemistry Laboratory, was the first to jump out from under the table, wearing only shorts and a green tie. He jumped up onto the table and declaimed "an amazing Shakespearean soliloquy about lysogeny" laced with humorous genetics jargon. "He performed it from memory, and he went on and on and on (I realized . . that it was probably largely *King Lear*, parts of which he had memorized as a schoolboy in South Africa)," recalls Bob Edgar, his lab partner for the course. "I have been in awe of Sydney ever since."[1]

At the time Edgar could not have predicted that he would one day work for Brenner and then see his lab partner, known for his unconventional behavior and clothing, standing one day, in a tuxedo, in front of the king of Sweden to accept the Nobel Prize in Physiology or Medicine for creating an entirely new field of developmental science.

Sydney Brenner's career in science was by no means linear, and his approach to the study of a tiny worm commonly found among decaying leaves took an equally circuitous route. Brenner was born in 1927 and raised in Germiston, South Africa, a small town on the outskirts of Johannesburg, where he lived with his family in the back rooms of his father's shoe repair shop. His Lithuanian immigrant father was illiterate, but Sydney, a precocious child, began reading at the age of four and teaching himself about science soon after. "The most interesting thing that I can recall from those days is discovering the public library," says Brenner. "There were no books at home of course, but I rapidly graduated to the adult library and read voraciously about lots of things."[2]

Books inspired his passion for biology, particularly the textbook *The Science of Life* by Herbert George Wells, Julian S. Huxley, and George Philip Wells.[3] Brenner was awed by the book's description of the ability of science "to draw the veil apart from nature." With insufficient funds to purchase the text, but unable to part with it, he eventually stole it from the library.[4] Brenner excelled in school, skipping several grades. At the age of 14, he received a scholarship to attend medical school at the University of the Witwatersrand in Johannesburg. "I was not a good medical student and had an erratic career, brilliant in some subjects, absolutely dismal in others," Brenner admits.[5] He was too young to practice medicine when he finished medical school and instead completed a master's thesis in cytogenetics. Several teachers and scientists who had recognized his low regard for clinical training but admired his exceptional skill at basic research helped him obtain a scholarship to Oxford University. In October 1952, he arrived at Oxford's Physical Chemistry Laboratory as a Ph.D. student to investigate how bacteria become resistant to viral attack. Brenner's graduate advisor, Cyril Hinshelwood, hypothesized that bacteria adapted to the viral attack, eventually protecting themselves from infections and destruction. Brenner thought that the bacteria change their genetic code or mutate as a form of resistance.

On a cold April morning in 1953, he drove from Oxford to Cambridge to examine the large model of DNA built by Francis Crick and James Watson. "The moment I saw the model and heard about the complementing base pairs I realized that it was the key to understanding all the problems in biology we had found intractable—it was the birth of molecular biology." While Brenner was examining the model of DNA, Watson and Crick were milling about the brick-lined room, both giddily talking about their discovery. This was the first time Brenner met Watson and Crick, and he later wrote that the afternoon was the "watershed" event of his scientific life. "The curtain had been lifted and everything was now clear." After viewing the model of DNA, Brenner and Watson went for a stroll around Cambridge. Brenner was excited about their finding, but at the same time deflated that his own bacteriophage research seemed "trivial" in light of the new discovery. Watson reassured him that his work was not inconsequential and had indeed put him "on the right road to enter this exciting new field."[6]

Brenner's studies on bacteriophages brought him to the Cold Spring

Harbor Course in 1954. Soon after completing his Ph.D., he began working as a biologist at Cambridge University, first at the Cavendish Laboratory and then at the Laboratory of Molecular Biology, where he shared an office with Francis Crick. Their desks were in the middle of the room, facing each other. This arrangement inevitably led to hours of conversation between the two scientists. Brenner remarked:

> The one thing that really characterized our conversations is that we never restrained ourselves in anything we said—even if it sounded completely stupid. We understood that just uttering something gets it out into the open and that someone else might pick up from that. There are people who will not say anything until they've got it all worked out. I think such people are missing the most important thrill about research—the social interaction, the companionship that comes from two people's minds playing on each other. And I think that's the most important thing. To say it, even if it's completely stupid![7]

In the next several years, Brenner made seminal discoveries, including the identification of mRNA. Brenner and Crick also determined how only three base pairs code for each of the 20 amino acids. All proteins are made of combinations of 20 amino acids. Since there are only four DNA bases (T, A, C, and G), a single base pair can come in only four combinations. A combination of two base pairs (4×4) provides only 16 combinations. Brenner proposed that it would take a minimum of 3 base pairs ($4 \times 4 \times 4$), or 64 possible combinations, to specify the 20 amino acids.

In October 1963, Sydney Brenner, now 36 years old, drafted a short proposal to Great Britain's Medical Research Council asking for the resources to launch a massive academic assault on a tiny worm. On a single page of paper, he drafted his plan to determine how intricate structures such as the nervous system are constructed from the instructions contained in DNA. Brenner's proposed research project caught many by surprise. "We think we have a good candidate in the form of a small nematode worm," he wrote.[8] Brenner thought that a teeny worm, no bigger than a comma on this page, made of only 969 cells, would be a simple starting point on the path to understanding more complex life. The worm, *Caenorhabditis elegans*, piqued Brenner's interest because it was one of the smallest animals to possess a nervous system—a combination of complexity and simplicity. After considering and rejecting most of the more typical experimental animals, including flies and protozoa, Brenner settled on this worm as the perfect laboratory candidate. With a voracious sexual, albeit hermaphroditic, appetite, it reproduces every three days and can be sustained on a simple diet of bacteria. Beginning with a single-cell embryo, he planned to map boldly all subsequent cell divisions—from egg to adult—until he could construct a detailed map showing exactly how to build a worm.

Brenner viewed the fruit fly *Drosophila melanogaster*, an established model, as too complex an animal for his purposes. He wanted a very simple organism that develops rapidly and is transparent so he could visually monitor cell division and migration. The worm seemed made to order. At the outset, many considered *Caenorhabditis elegans* to be a "joke organism,"[9] regarding it as too simplistic to warrant such an enormous effort. But with Brenner's successful track record, the Medical Research Council took a risk and provided funding to initiate the worm project.

Brenner's first step was to put together a team. There were almost no previous studies to guide researchers, on either the worm itself or the kind of mapping Brenner hoped to accomplish. This challenge attracted a unique and eclectic crew of scientists from many backgrounds and nationalities. "People asked us what the qualifications were for them, and we said, 'Just interest in the subject!'" Brenner recalled.[10] The recruits started coming. One of the first was John Sulston, an organic chemist who loved hands-on lab work (what he refers to as "playing with toys") and almost abandoned his chemistry studies at Cambridge to pursue a career in theatrical lighting. When Sulston joined the group in 1969, he was 27 years old and looked like the stereotypical hippie with an overgrown beard and long hair, usually wearing "Jesus sandals," but known for his quiet demeanor and keen insight.[11] Before joining the group, Sulston recalls, "There were lots of jokes about Sydney's worm, and general skepticism about its chances of coming to anything. This seemed a pretty good recommendation to me: there's little point in doing what everybody else is doing."[12]

In November 1974, another unlikely recruit, Robert Horvitz, a wiry and intense Chicago native with thick glasses and thicker sideburns, joined the worm group. Horvitz was fresh out of James Watson's laboratory at Harvard University and, like Brenner, his doctorate work showed how a virus modified an infected bacterial cell. Before accepting the position, Horvitz asked several people about "Brenner and his nematodes," amused that the phrase sounded "like a new rock band."[13] Brenner's group received rave reviews. Horvitz arrived in Cambridge, and he and Sulston immediately became close friends.

As the laboratory began to expand and papers concerning the worm appeared in the scientific literature, developmental biologists started to become more familiar with *Caenorhabditis elegans*. Researchers who worked in Brenner's laboratory began fanning out across the globe, each taking frozen stocks of worms and colonizing new research labs. Brenner's former Cold Spring Harbor lab partner, Bob Edgar, also visited Brenner's Cambridge laboratory. Upon returning to his professorship at the University of California, Santa Cruz, Edgar further solidified the informal network by organizing international worm meetings and starting a newsletter, *The Worm Breeder's Gazette*.[14] The first issue appeared in December 1975 and consisted mainly of articles on how to tend worms. For example an article by Edgar, "The Shipping and Handling of Nematodes," was simply a short paragraph on how to mail worms in letters. "I have had some success shipping worms in letters. Three or four small pieces of filter paper are placed on a piece of aluminum foil. The filter papers are saturated with a buffer washate from a starved plate. The foil is folded several times to create a seal. The recipient places the filter papers on a plate or washes them in buffer."[15]

In 1977, Horvitz and Sulston published their studies of the development of the worm and the cell lineage of the postembryonic stage.[16] They discovered that many more *Caenorhabditis elegans* cells were produced during embryogenesis than managed to survive, which unequivocally showed that certain cells were programmed to die as a normal part of development. The worm generates 1,090 cells during its development, but the adult animal has only 969 cells. Therefore, 131 are slated to die during maturation. It was unclear why the worm had this seemingly wasted development. Several genes were later identified that regulated the process of programmed cell death. Once those genes were activated, the cells died on command—in effect, by committing suicide. Later in his career, Horvitz learned that one of the genes that controls cell suicide in nematodes is almost identical to a gene found in humans. This gene plays a critical role in the development of cancer. When the reproduction of a cell becomes aberrant, it initiates a programmed cell death routine. When this gene does not function correctly, cancerous cells do not commit suicide and instead proliferate.

On a trip back to Boston in 1977, Horvitz bumped into Martin

The Worm Breeder's Gazette

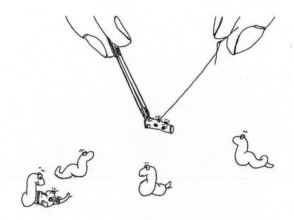

The Worm Jean Sequins Project

Volume 13 No. 1

October 1, 1993

The Worm Breeder's Gazette. Drawing by William G. Wadsworth.

Chalfie, an old school friend.[17] Chalfie was completing his Ph.D. in neurobiology at Harvard University and the two spent an afternoon catching up. Horvitz and Chalfie both attended Niles East High School in Skokie, Illinois. In talking with Horvitz, Chalfie was impressed with the atmosphere in Brenner's laboratory and the success that Horvitz had had in completing the cell lineage studies of *Caenorhabditis elegans*.

"I wonder if Sydney might take me as a postdoc?" asked Chalfie. This short conversation inspired him to write to Brenner about job opportunities, and Brenner replied that he would be happy to have him. Chalfie began as the first neurobiologist to work on the worm. In 1977, a few months before Chalfie was scheduled to ship off to England, the first international worm meeting was held at Woods Hole and Horvitz and the rest of the worm crew came to the United States. In the car on the way to the meeting Horvitz suggested to Chalfie that he speak to Sulston about the cell lineage project. In particular, Sulston had found cells thought to be touch receptors, but had not had time before the meeting at Woods Hole to pursue the topic further. At the meeting, Chalfie approached Sulston, who was presenting his work in a poster session at the meeting. During these sessions, researchers stand next to a poster displaying their data. But instead of having his data printed on a large poster, Sulston was standing in front of a window with 35-millimeter projector slides taped to the glass. He thought he was giving a slide presentation. Chalfie squinted to examine the slides. Fascinated by the topic, he decided he would like to pick up the project. Later he remarked that mechanosensation became the major focus of his scientific career.

The Brenner laboratory was like a well-oiled machine in those days. The power of this newly minted model allowed new laboratory members to plug into the process, pick up the techniques, and start generating interesting data almost immediately. Chalfie was no exception. After spending only four years in Brenner's lab, he had collected an impressive set of data and was offered, and accepted, an assistant professorship at Columbia University in New York. In his new position Chalfie continued to approach science as a creative endeavor. The cell lineage studies of Brenner, Horvitz, and Sulston had demonstrated a valuable property of *Caenorhabditis elegans*—its small size and its transparency. Researchers had taken advantage of these characteristics to observe living specimens directly under a light microscope. Standard light microscopy can penetrate only a few cell layers. In mammals this amounts to just the superficial layers of the skin, but an entire *C. elegans* is only a few cell layers thick. A researcher could see the animal's nerve cells, muscle cells, and digestive tract just by looking through a microscope. But although the small size and transparency of *C. elegans* make it well suited to visualizing living cells, it was not possible to see the cells' proteins and DNA. These molecules are too small to be seen with standard microscopes. It was not possible to use Coons's technique of immunofluorescence in the worm because the process of coaxing the relatively large antibodies into the animal's cells inevitably was lethal to the animal.

In May 1988, Martin Chalfie attended a weekly informal talk in Columbia's biology department called Neurolunch. Paul Brehm, a researcher from Tufts University, was discussing his work on the marine hydroid *Obelia geniculata*, a bioluminescent organism that resembles an old-fashioned quill pen. Bioluminescence was far outside Chalfie's own research, but he listened patiently while eating his lunch. Fifteen minutes into his talk, Brehm mentioned a strange feature of the bioluminescence system in this *Obelia*: it appeared that in addition to producing light this animal, like the jellyfish *Aequorea*, had a green fluorescent protein to convert the blue light into green light. "I nearly fell out of my chair," Chalfie recalled. He was astonished by the potential of a fluorescent protein for worm research. Chalfie says that he got so excited that he didn't listen to another word of the Brehm's talk. "It's in large part because of the *C. elegans* work that I was interested. I honestly believe that if I was listening and I was working on mice or even *Drosophila*, I would have not been interested."

After the talk, Chalfie and Brehm had a conversation in Chalfie's laboratory. Chalfie asked, "Could this [green fluorescent protein] be expressed in another organism?" What he really wanted to know was whether the DNA for the fluorescent protein could be incorporated into *C. elegans*. This feat would swindle the animal into synthesizing a fluorescent protein, producing a miniature lantern when certain genes are read. If this could be achieved, it would be possible to observe, in living animals, when and where genes are flipped on and off. Since Chalfie was working with a thin and almost "see-through" animal, this would be an especially valuable tool. From his previous research, he also knew that when specific worm genes were inactivated, the animal lost its ability to respond to touch. Normally, *C. elegans* defensively reacts when gently prodded with a toothpick, but if a mutation was made to these genes, the worm failed to react when poked. Chalfie hoped to illuminate these "touch" genes to see where they are located and what function they played.

Transferring genes from one animal to another is an art form as well as a science. When extending beyond bacteria to more complex creatures, it becomes increasingly difficult to modify and manipulate an organism's DNA. In order to slip the fluorescent protein gene into *C. elegans*, for example, embryos are first injected with the jellyfish DNA through a small glass pipette. The fluorescent protein DNA then slinks into the cell's nucleus, where it is read by the cell's machinery, instructing it to produce fluorescent proteins. When the animal reproduces, the glowing gene is also passed along to its offspring. This is similar to how other transgenic animals are made. It differs, however, from animal cloning in that only a small piece of DNA is added, rather than the entire genome swapped.

At the time, Chalfie was not following the fluorescent protein field closely. This worked in Chalfie's favor because the overwhelming consensus held that the protein would not fluoresce on its own. All the major researchers who had worked with GFP, including Osamu Shimomura, Bill Ward, and Doug Prasher, felt it would not fluoresce if produced in an animal other than the jellyfish. Green fluorescent protein was not the first fluorescent protein to be pursued as a molecular marker. Several other proteins were known to produce visible light fluorescence, most notably the phycobiloproteins. Isolated from certain strains of photosynthetic cyanobacteria, these proteins transfer absorbed light to the light-harvesting proteins. The sequences of these proteins had been known for years. Like GFP, these proteins were highly fluorescent when illuminated with specific wavelengths of light. Analysis of these phycobiloproteins, however, revealed that their fluorescence arose from chemicals added to the protein after synthesis, and not from the protein itself. Bacterial enzymes added complex fluorophores to the outside of the proteins—causing them to be fluorescent. These fluorophores are produced in the bacteria by a synthetic pathway requiring a number of bacterial-specific enzymes. If a researcher transfers the cDNA encoding the phycobiloproteins into another organism, the proteins can be produced, but they will not be fluorescent. In order to make the completed fluorescent protein, a researcher would have to transfer all the genes in the catalytic pathway that produce the fluorophore and the enzymes that attach these to the protein. This is not a very practical way to make a worm cell fluorescent. (At the time, it was assumed, incorrectly, that green fluorescent protein required enzymes to produce its fluorescence.)

Chalfie spent the rest of the day after Brehm's talk probing the scientific community, looking for someone working on the molecular biology or biochemistry of GFP. He eventually came upon Douglas Prasher, who at the time was at the Woods Hole Oceanographic Institution. Chalfie heard that Prasher was trying to isolate the gene for GFP. Chalfie phoned Prasher, who said that he was close to obtaining the full cDNA sequence, and he would call Chalfie when he had finished. "This is terrific," replied Chalfie. "We have a transparent organism to put it in. We have cell-specific promoters to drive it. This is going to be lot of fun." Chalfie never received a return call.

In 1991, Prasher was working on GFP only a few buildings away from Shimomura, who was now at the Marine Biological Laboratories, also in Woods Hole. Oddly the two never collaborated and had only one interaction. "I wasn't the type to go up and talk to him," says Prasher. "I just wouldn't do it." One time while both were at Friday Harbor collecting jellyfish, Prasher gave a short talk on his effort to determine the sequence of GFP. "That was the only time I was ever in the same room," says Prasher. After the talk, they briefly conversed. "He didn't think if it expressed, it would be fluorescent," Prasher recalls.[18]

For several years, Prasher continued his lonely quest to determine the genetic sequence of GFP. When Prasher finally completed the task, there was no celebration.[19] His funding from the American Cancer Society had run dry almost 2 years earlier and he was working by himself at an institute that he says had "very little interest" in his research. "I was convinced that writing grants and working in an isolated environment was not my cup of tea," says Prasher. A job became available in a U.S. Department of Agriculture facility at Otis Air National Guard Base studying gypsy moths, which Prasher dismissed at first: "I had a horrible attitude about working for the government." But then he reconsidered: "I wouldn't have to relocate. I wouldn't have to write grants anymore. It sounded pretty good." So following publication of the GFP sequence, in July 1992, Prasher left behind Woods Hole, and fluorescent protein research, forever.

Later in 1992, Ghia Euskirchen, a first-year doctoral student at Columbia University, approached Chalfie about completing a rotation in his laboratory. New graduate students at Columbia spend a few weeks in different rotations before settling in the laboratory where they feel most comfortable. Euskirchen had already completed an M.S. in chemical engineering and had worked with fluorescent compounds. Euskirchen's qualifications led Chalfie to recall his conversations with Brehm years earlier. He sat down at the computer with her and typed "fluorescent protein" into MedLine, the free national scientific database that had just recently been installed at Columbia. Chalfie recalled: "All of a sudden it comes up with Doug Prasher's 1992 paper and I said, 'My god, he finished it. He did it; why didn't he get in touch with me?' We ran downstairs to the library, got the thing; it had his phone number in it. We called him up."

Prasher was happy to ship off the cDNA clone and Euskirchen began her research. She used the recently developed polymerase chain reaction technique to subclone the GFP gene from the plasmid Prasher had sent to them and placed it into another plasmid that allowed the protein to be made in bacteria. After causing the bacteria to take up the green fluorescent protein's DNA, Euskirchen put some of the bacteria on a slide and examined them under Chalfie's low-quality fluorescent microscope. She didn't see anything. Just as a double check, she went to her former laboratory and looked at it through a more sophisticated microscope. She was surprised to see brightly fluorescent glowing bacterial cells spread about the slide. Her experiment had worked. Despite a longstanding belief to the contrary, the DNA for green fluorescent protein could be extracted from jellyfish and made to work in other organisms. Twenty-six-year-old Ghia Euskirchen was the first to see the glow that would change the face of biological science. Yet, despite her stunning success, further research with GFP did not interest Euskirchen, and she left Chalfie's laboratory after her short rotation.

For the next year, Chalfie was left working on his own, trying to get the protein to fluoresce in his favorite organism, *C. elegans*. Chalfie was interested in identifying four specific cells out of the 969 that make up the adult worm. These cells are neurons that sense and convey touch sensation in the animal. Chalfie already knew that these neurons specifically express a very high level of the protein coded by a gene called *mec-7*. When this gene was disrupted, the worm no longer responded to being touched. Chalfie placed a portion of the promoter region of the *mec-7* gene in front of the green fluorescent protein sequence in a plasmid. This plasmid was injected into adult worms and taken up into the worm's own DNA. When offspring of the injected worms were examined, they showed strong green fluorescence only in the four touch neurons. The neurons and their long appendages were fluorescently labeled. This labeling approach allowed Chalfie to watch the worm as it grew and see exactly when the *mec-7* gene was turned on and when the cells obtained their neuronal personality.

In the October 1, 1993, issue of *The Worm Breeder's Gazette*, Chalfie published a five-paragraph article called "Glow Worms: A New Method of Looking at *C. elegans* Gene Expression." The article begins:

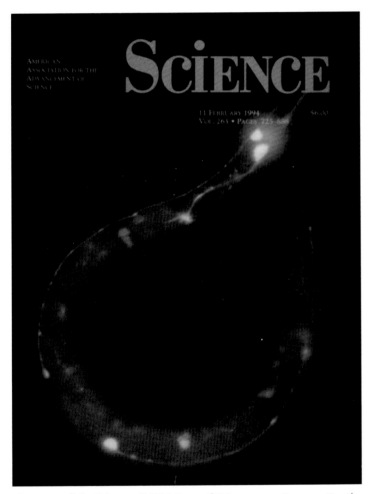

The cover of the February 11, 1994, issue of *Science* magazine reporting the expression of green fluorescent protein in bacteria and the nematode worm *C. elegans*. Photo by Martin Chalfie; reprinted with permission from *Science*, vol. 263, February 11, 1994; © 1994 American Association for the Advancement of Science, Washington, D.C.

We have developed a new way to look at gene expression in *C. elegans* (and other organisms) that utilizes an inherently fluorescent protein (the green-fluorescent protein; GFP) from the jellyfish *Aequorea victoria*. GFP fluoresces bright green when illuminated with blue light. We have found that this fluorescence does not depend upon any other component specific to *A. victoria*, so GFP can be used instead of lacZ, for example, to make gene expression fusions.

He concluded the article as follows: "We have lots of ideas of how GFP might be used and imagine that other people will have many more . . . We have generated a set of plasmids that may be useful for *C. elegans* researchers . . . If you are interested in obtaining these clones, please write (or FAX or email) your request."[20]

Four months later, on February 11, 1994, the article "Green Fluorescent Protein as a Marker for Gene Expression" appeared in the prestigious journal *Science*.[21] Chalfie had the courage, motivation, and skill to test the idea that green fluorescent protein could be expressed and fluoresce outside of the jellyfish, and it paid off. The protein was almost tailor-made for the new molecular revolution. New methods of cell culturing and transfection allowed the expression of proteins in cultured mammalian cells, and transgenic techniques allowed the expression of foreign genes in almost any organism. Within a few years green fluorescent protein would be expressed in plants, frogs, fish, mice, goats, rabbits, monkeys, flies, beetles, lampreys, and yeast. The protein is highly suitable for expression in almost any organism, but the story was only beginning. Within a few years, green fluorescent protein would be completely reinvented.

CHAPTER EIGHT

Fluorescent Spies

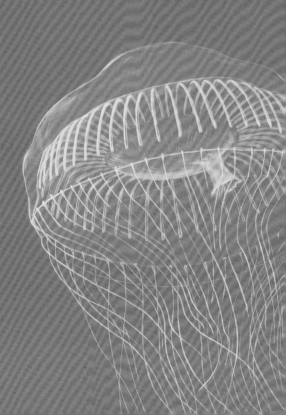

NESTLED in the sun-drenched coastal campus of the University of California at San Diego is a laboratory run by one of the world's leaders in molecular espionage. For the majority of his prolific career, Roger Yonchien Tsien has been devising techniques to spy on how cells function. Tsien's optical molecular probes have reinvented how biologists study the secret lives of cells. He has written hundreds of scientific articles, has garnered scores of professional awards, and holds over 60 U.S. patents. Tsien's ideas flow abundantly and profitably, one leading to a billion-dollar biotechnology company that he cofounded in 1995. But Tsien doesn't seem to be driven by wealth. He prefers to dress in button-down denim shirts and to bicycle to work, and he can often be seen carrying his partially blind and paralyzed dog Kiri around the streets of La Jolla. Members of his laboratory say he appears most comfortable pacing around his large third-floor laboratory, pondering ways of pushing forth his research and inventing new ways of studying biology.

Tsien views the human body as a "medium-sized city."[1] He doesn't see it as a metropolis, since, as he puts it, the human genome contains *only* 35,000 genes, not many more than a sprawling weed has—the mustard plant, for example, has 27,000 genes. To Tsien, the Human Genome Project produced a useful citizen directory, but it didn't provide insight into what he finds most interesting: how the inhabitants live and carry out business. "Like town dwellers, individual protein molecules in a cell are born, get modified or 'educated,' travel around, and cooperate or compete with each other for partnerships. Some proteins emigrate from the cell. A few have the job of killing other proteins," said Tsien in a speech delivered after he won the Dr. H. P. Heineken Prize for Biochemistry and Biophysics.[2] His research has been primarily focused on understanding this cellular "anthropology." Using chemical and molecular techniques he is able to observe and report on the social interactions of proteins. Tsien has also developed more sophisticated espionage systems whereby he can attach a molecular "radio collar" to assassinate a protein to see which cell activities continue or collapse as a result.[3] The exploding radio collar works by a process called chromophore-assisted light inactivation. When such probes are subjected to light, radical oxygen molecules are created that kill almost all tagged proteins in less than half a minute. Tsien communicates with his probes not via radio waves but by fluoresced light.

When Chalfie's glowing worm article appeared in *Science*, it was the first time most of the scientific community had heard of green fluorescent protein. But Roger Tsien had for some time been trying to find a way to enlist fluorescent molecules into his spy corps. Almost immediately after Prasher published the sequence of the jellyfish's fluorescent protein in 1992, Roger Tsien contacted him, asking for a sample of his clone. Prasher, then still at the Woods Hole Oceanographic Institution, was willing to send out samples of the cDNA, and told Tsien that he had run out of funding, was abandoning his lab, and planned to leave the field completely and work on mosquitoes at a nearby USDA facility on Otis Air

National Guard Base. In fact, only two people contacted Prasher requesting the green fluorescent protein gene: Martin Chalfie and Roger Tsien.

The prospect of inserting the jellyfish's fluorescent protein into any cells greatly intrigued Tsien. If it worked, it would behave like a native-born spy—one born directly into the chosen cell; it would not have to be forced to pierce the cell's defenses. For two years before Chalfie's *Science* paper appeared, Tsien was racing to organize his laboratory so he could express and reengineer the protein—an area of science where he had little expertise. But with experience in organic chemistry, biophysics, and biochemistry, he wasn't at all hesitant to take on an additional discipline and expand into molecular biology.

Born in New York City on February 1, 1952, Tsien was raised in a middle-class house in Livingston, New Jersey. His parents came to the United States from China in the 1930s so his father could study engineering. As an adolescent Tsien became fascinated with chemistry sets, which he played with in his basement, but he quickly became bored with the many safety features of the store-bought sets. "I found an old-fashioned chemistry text somewhere in one of the school libraries that actually had some much neater reactions with much more dangerous chemicals," says Tsien.[4] Soon he progressed to making gunpowder. On one occasion, Tsien and his two brothers used the gunpowder to make a homemade grenade out of "aluminum-foil-chicken-pie-TV-dinner-dishes." The grenade failed to explode, but it did briefly set a portion of the Ping-Pong table ablaze and fill the house with smoke.

As a senior at Livingston High School, Tsien won first place in the 1968 Westinghouse Science Talent Search, America's oldest and most prestigious high school science competition, with the winner often called the recipient of the junior Nobel Prize. His project, one of thousands of entries, was modestly titled "Bridge Orientation in Transition-Metal Thiocyanate Complexes." He began the research during a pre-college summer program at Ohio State University and later completed it on weekends at Columbia University after "inveigling" professors to let him use spare laboratory space. The Westinghouse award provided Tsien with a scholarship to attend Harvard University, where he majored in chemistry and physics. He was placed in a special program for "hot shots," but the late 1960s were a turbulent time at Harvard, and Tsien did not approve of the rigidly taught chemistry classes. With numerous advanced placement chemistry credits from high school and a strong interest in playing the piano, he was able to complete his science major while taking as many music classes as chemistry classes. During his last year at Harvard, Tsien became interested in fluorescence and began thinking how he could devise visual techniques to study the brain.

In January 1972, Tsien received a letter from the Marshall Aid Commemoration Commission offering him a full scholarship to study at Cambridge University. The letter said that the commission had appointed Richard Adrian, a skeletal muscle physiologist, to be his advisor. Tsien was livid. He had no interest in muscle. Immediately, he called his older brother, Richard, for advice. Richard, a prominent neurophysiologist, was a Rhodes Scholar at Oxford, and Roger knew that Adrian had served on his brother's thesis committee. Tsien complained to his brother: "I don't want to work on skeletal muscle. Muscle is a backwater. I want to study the brain." His brother replied: "Don't worry, Richard Adrian is a true British gentleman and will let you work on whatever you really want. They are much looser over there." Roger replied: "In that case, I'll just keep my mouth shut, I won't write a letter of protest to the Marshall Commission. I'll see how it goes."

Once Tsien arrived at the Physiology Department at Cambridge University, he became further obsessed with studying the living brain. His presumption was correct: Richard Adrian was a poor advisor match. Adrian couldn't provide guidance to Tsien, who was mainly interested in using fluorescent dyes to visualize the brain. At Cambridge, Tsien further realized he didn't want to study the brain using the established and widely used methods such as "drilling a hole in the skull and dropping in electrodes." Tsien began working by himself in an undergraduate teaching laboratory in the Department of Chemistry, where he was considered a "carpetbagger," without an advisor in the Chemistry Department. "I was alone in this large echoing room that was occasionally busy during class time, but the majority of the time, it was just empty."

His first big success came after he completed his doctorate and began working in Timothy Rink's laboratory at Cambridge in 1978. There he developed one of the first fluorescent reporters, dubbed quin-2, which quantifies the presence of free calcium (Ca^{2+}) ions inside cells. In the presence of calcium ions the dye changes its fluorescence intensity. Although calcium is very abundant outside cells, there are very few free calcium ions inside of them. Cells use the brief increases in calcium ions as a signaling mechanism. In the same way the jellyfish *Aequorea* uses transient increases in calcium to produce light from Aequorin, cells throughout the body use calcium to activate processes such as muscle contraction, neurotransmitter release, cell division, and insulin secretion. When a cell is stimulated, channels in the cell's membrane open and allow a flood of calcium ions to enter the cell. These free calcium ions then bind to a range of proteins, snapping them into action. Soon after this flood, the free calcium is rapidly cleared by a variety of mechanisms. This process can only be studied in living cells. Tsien sought a way to convert the invisible movements of calcium ions into a process that could be observed under a microscope.

The fluorescent dye quin-2 has a pocket that is just big enough to catch the Ca^{2+}, but too small to capture the slightly larger, and more abundant, free magnesium (Mg^{2+}) inside the cell. The dye is injected into cells and, when the levels of Ca^{2+} are elevated in the cell, the free calcium sticks in the pocket. This changes the dye's conformation, which increases its level of fluorescence. By simply monitoring the cell's fluorescence intensity, it is possible to measure the amount of free calcium inside the cell. Tsien's dye provided a glimpse of the complex and fluid way in which calcium levels change in cells. Names like "calcium sparks" and "waves" were coined to describe the newfound phenomenon. The dyes produced images that transformed the static science of biochemistry into a dynamic and highly visual discipline.

In 1981, Tsien left Cambridge and accepted an assistant professorship at the University of California at Berkeley. He was then 29 years old and had 12 papers under his belt, including several published in top journals that described his breakthrough nondisruptive technique for rapidly measuring free calcium in living cells. At Berkeley he developed an even better calcium dye, called fura-2, that offered a fluorescent signal 30 times greater than his previous version (already widely used at the time). The new dye became so popular that the paper initially describing its characteristics has been cited over 15,000 times, and it is among the five most commonly cited papers in science over the past 20 years.[5] The Institute for Scientific Information, which tracks citations, considers an article a "classic" when it receives 1,000 citations.

While Tsien was at Berkeley, a fellow faculty member, Alexander Glazer cloned the first genes of phycobiliproteins from photosynthetic cyanobacteria. Phycobiliproteins help capture sunlight and funnel its energy to the cyanobacteria's photosynthetic apparatus. In a brief phone conversation, Glazer told Tsien that when he mixed the cloned phyco-biliprotein with some of the cyanobacteria's original pigment, he got fluorescence. "Could you try this in another cell type?" Glazer asked off-handedly. Tsien sat in his office and thought about the brief conversation. An hour later, he called Glazer back. "Alex do you realize what you have here? If you could get this to work, you could track any protein in the cell by just fusing it to the phycobiliprotein and add pigments from outside. That would be a gold mine."

Glazer did not seem interested in technological applications and was mainly focused on the intrinsic properties of the protein. "I got the impression that this greed of mine had not occurred to him," says Tsien. Phycobiliproteins have essentially been dropped as molecular spy candidates since they are bulky molecules that require two other enzymes in order to fluoresce. "The real system would eventually have to have three genes in it—the fundamental protein and two enzymes. That scared me. That really scared me," said Tsien. But the wheels were turning in Tsien's mind and he began thinking of ways he could program a cell to genetically manufacture its own fluorescent signals rather than having to insert synthetic molecules.

In 1989, Tsien moved to the University of California at San Diego. He saw this relocation as a perfect opportunity to distance himself from his calcium probes and expand into new areas. He began working with Susan Taylor, who headed a laboratory in the same building, on ways to measure cyclic adenosine monophosphate (AMP) in a cell. Cyclic AMP, like calcium, is an intracellular messenger that regulates various cell processes and mediates the effects of many hormones by relaying signals arriving on the cell surface to proteins within the cell. Working with Taylor, Tsien developed a successful, yet tedious, method of visualizing cyclic AMP levels in living cells. This required taking a lot of protein, labeling it in test tubes with fluorescent dyes, purifying it, putting it back together, and microinjecting that into cells. "That was working, but it was very clumsy and limiting. We felt the need to have a molecular biological way of encoding fluorescence rather than taking organic chemical dyes and attaching them onto proteins," Tsien recalls.

Late one afternoon in May 1992, Tsien was sitting in his office thinking of a way to improve the process of labeling proteins with dyes. The University of California had recently obtained access to Medline, and, as Martin Chalfie had done, Tsien typed in the words "fluorescent protein." On the screen appeared Prasher's article, "Primary Structure of the *Aequorea victoria* Green-fluorescent Protein," which had just been published in *Gene*.[6] Tsien saw that the article included Prasher's phone number. The next morning he telephoned him. Tsien recalls:

> Doug was willing to share DNA on condition that if we got anywhere with it we would make him a coauthor, which I said was perfectly fair. To my astonishment, he wasn't going to work on it anymore. I could see all these potentials for it. He did tell me not to get my hopes up too high. He had tried to express it in *E. coli* . . . and it had not become fluorescent. This was the full-length sequence at last and he hadn't tried it. As I recall, I asked if anyone else was working on it. He said, "no." I had the impression that I was the first person to have noticed the paper practically. At least noticed it enough to phone him up. He promised to give the DNA, but I didn't ask for it right away, because I had nobody to work with it. We had not done any molecular biology in the lab. It wasn't our focus.

In late September 1992, Roger Heim, a soft-spoken triathlete and newly minted Ph.D. from the Swiss Federal Institute of Technology, joined Tsien's laboratory. Heim arrived with the intention of learning cell imaging techniques. But Tsien recognized that he was the only person in the laboratory with training in recombinant DNA technology and immediately assigned Heim to the fluorescent protein project. At that point, Tsien placed a call to Prasher and said, "We're ready, send the DNA." Prasher agreed to comply, but to Tsien's disappointment, he also mentioned that he had already shipped samples to Martin Chalfie at Columbia University. Tsien, recognizing the potential advantages of expressing the protein, felt great pressure to demonstrate that the jellyfish protein could be inserted into another organism and still fluoresce. Heim set to work. With a yeast specialist, Scott Emr, in his building, Tsien decided they should work to express the jellyfish protein in yeast because Emr could provide expertise. But, shortly after the race to express the protein began, the competition was over. Three weeks after beginning work, Chalfie mentioned to Tsien that he had expressed green fluorescent protein in bacteria. Tsien recalls: "Was it a disappointment? Maybe very very slightly, but to be honest, we had just barely got started and he had already gotten fluorescence. It wasn't like we had already run a long race and then got beaten just before the finish line."

Although Tsien and Heim were defeated in their attempt to become the first people to express the jellyfish's fluorescent protein in another organism, they were elated to know that the protein could fluoresce by itself, without other jellyfish enzymes. Tsien had previously feared that green fluorescent protein would require accessory enzymes, such as those needed by phycobiliproteins, to fluoresce. Now he knew it was *possible* to enlist the protein as a self-sufficient spy.

Nevertheless, it took Tsien and Heim five additional months to achieve any success expressing the protein. Chalfie had judiciously subcloned the green fluorescent protein in such a way as to remove a suppressing part of the sequence that resided in front of the coding region of the gene. In March 1993, Tsien and Heim had a minor success: they achieved sporadic green fluorescent protein production in yeast. "Maybe

1 in 100 was fluorescent," says Tsien. In another phone call, Chalfie told Tsien that he was no longer working with bacteria and instead was trying to express the fluorescent protein in his model organism, the worm *C. elegans*. Tsien was amazed that Chalfie was going to refrain from publishing his finding that green fluorescent protein could be expressed in another organism until he got it to work in the worm.

"Marty was obsessed with not just giving a technical demonstration. He was a hard-core biologist who felt he had to prove that it was going to be useful and teach you something new about biology," says Tsien. "There is a very strong ethos in most of the biological community that this stuff is just technology. And you have to learn something new about intrinsic biology; only intrinsic biology is pure and worthwhile and all the rest is just techniques." Tsien believes that science for science's sake is self-indulgent, especially when the science can be applied for the betterment of humanity, or even the potentiality of profit. He believes the development of techniques is an equally noble science. "So . . . what was so important about that particular worm thing? It was such a technical breakthrough that I doubt there is one person in 500 who uses green fluorescent protein that can say what it was that Marty actually showed in worms," says Tsien.

Tsien says he was tempted to try to rush into print with his own findings. But he didn't because he thought that would be "ambulance chasing" since he didn't have new information to add. The way Tsien looked at it, Chalfie was the first person to express green fluorescent protein.

A researcher who had worked with Edmund Newton Harvey, Frederick Tsuji, and a coauthor published a paper in February 1994 in a lesser-known journal, *FEBS Letters,* titled "*Aequorea* Green Fluorescent Protein: Expression of the Gene and Fluorescence Characteristics of the Recombinant Protein."[7] This paper appeared a few weeks after Chalfie's paper.

Tsuji, 70 years old at the time, had just returned to the United States from a six-year stay in Japan, where he had also cloned the jellyfish protein Aequorin. Conceptually, this paper was essentially the same as Chalfie's paper, but it received little attention and little recognition.

After Chalfie published that GFP could be expressed in *C. elegans* on February 11, 1994, he retreated from the study of fluorescent proteins and went back to his work on mechanosensation. Word began spreading throughout the scientific community of the ease with which green fluorescent protein could be expressed. Chalfie received hundreds of requests to share the DNA coding for GFP, and he routinely mailed it to laboratories around the world. But although the jellyfish's green fluorescent protein proved to be a very valuable tool, its fluorescence was not brilliant. The jellyfish protein absorbed light over a broad range of colors. That made it difficult to get maximum fluorescent output by exciting the protein with only one color of light. In practice, this property makes the protein's fluorescence weak. Tsien and other scientists immediately recognized that the jellyfish protein could possibly be engineered to glow brighter and have a sharper excitation peak. A new pursuit ensued to tinker with what Tsien referred to as the jellyfish's "dim, fickle, and spectrally impure" fluorescence. Researchers set out to soup it up, like a race car, producing a supercharged fluorescent protein.

The unique feature of green fluorescent protein is that three amino acids in the protein's peptide backbone spontaneously form a fluorophore by a series of self-induced chemical reactions. At that time, it was unclear how or why this happened. Researchers did know, however, that the three-amino acid fluorophore wasn't enough to produce fluorescence. Scientists had chemically synthesized just the fluorophore and it didn't fluoresce. Unknown aspects of the protein enabled the fluorescence. Tsien assumed that if he altered the sequence of the protein, he could alter its properties. By planned and random techniques, Heim, Tsien, and other scientists swapped virtually every amino acid in the 238 amino acids of the jellyfish protein. To check for fluorescence, Heim used an old spectrofluorometer, referred to as the Green Monster. The operation was low-tech. Heim looked through different colored Kodak filters that he taped to his lab goggles while changing the color light by hand on the old Green Monster. "At Berkeley they were ready to throw it away, but as an old pack rat or magpie, I refused to let it go," Tsien says of the spectrofluorometer.

In most cases, mutations either had no obvious effect or disrupted the formation of the fluorophore, extinguishing the fluorescence. But one day Heim noticed a single bacterial colony that appeared to be glowing blue. He picked the colony off the plate and examined it further. Heim, Prasher, and Tsien published the mutation for blue fluorescent protein in December 1994.[8] Then Tsien decided to deliberately mutate the first amino acid in the 3-amino acid fluorophore. He hit the jackpot. It was a single amino acid substitution that created this supercharged fluorescent protein, a substitution of a therione for a serine at position 65 of the protein's sequence, a S65T mutation. In mutation terminology, the letter of the original amino acid to be mutated is the first letter. That letter is followed by the position of the amino acid in the protein sequence (serine is the 65th amino acid of green fluorescent protein's 238 total amino acids). The last letter represents the amino acid that is substituted for the original. This tiny change of one amino acid resulted in a dramatic difference in the spectral characteristics of the protein. No longer did the protein have two humps in its absorption spectrum. With this change, the protein only excited at one peak color, at 488 nanometers. Tsien's group published these results as "Improved Green Fluorescence" on February 23, 1995, in the journal *Nature*.[9]

Tsien called this mutant of GFP enhanced GFP, or simply eGFP. This mutant had a less complicated excitation spectrum and was eight times

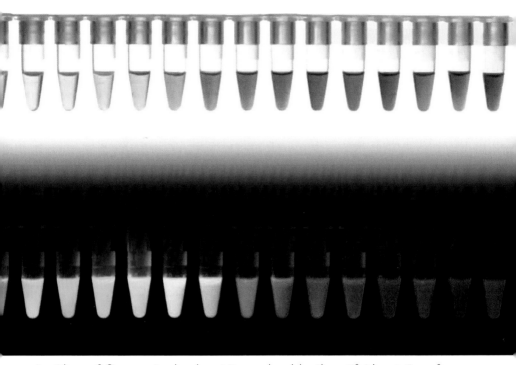

A rainbow of fluorescent-colored proteins produced by the artificial mutation of jellyfish green fluorescent protein. Alterations in the amino acid sequence of the natural protein shift the excitation and emission wavelengths. Photo by Roger Y. Tsien.

brighter, a decided improvement for scientific use. Having much experience with calcium dye development, Tsien included this innovation under a broad patent entitled Modified Green Fluorescent Protein that would cover most mutagenic modifications of GFP and secure strong intellectual property rights.[10] Within a year of the appearance of Chalfie's paper, several other laboratories published a range of mutations in green

Roger Tsien holding a rack of vials containing fluorescent proteins his laboratory created. Photo by Joe Toreno for the Howard Hughes Medical Institute.

fluorescent protein that produced different effects on the protein's properties. Some changes produced brighter proteins, some more stable proteins, some proteins less affected by pH changes, and still other researchers reported that specific mutations caused the protein to form better at mammalian body temperature (98°F) than it did under its natural conditions (around 40°F).

All this work was done without anybody knowing what the molecule actually looked like. Tsien knew that understanding the structure of the protein would allow him to alter its properties, improving its fluorescence. The most precise way to determine the detailed structure of a protein is via crystallography. It was, after all, the crystal diffraction data of DNA provided by Rosalind Franklin and Maurice Wilkins that gave Watson and Crick the foundation for proposing the double helix structure of DNA.

Because single protein molecules are too small to interfere with visible light in a way to form an image, electromagnetic radiation of a shorter wavelength must be used. X-rays have a much shorter wavelength but are invisible to the human eye. X-ray crystal diffraction is a method that involves focusing a very strong beam of X-rays onto a highly pure crystal of a protein. When the X-ray interacts with the crystal's repeating structure, it emerges from the crystal and forms a diffraction pattern. This pattern is not an image of the individual proteins but rather a circular arrangement of small dots. As the X-ray beams pass through the crystal, they bend in repeatable ways based on the repetitious structures in the crystal. The pattern represents a list of distances between various aspects of the protein. The procedure is like transforming an image of a house into a list of measurements. The difficulty is taking this list and determining which measurements are which and what is the starting orientation. Within the diffraction pattern, the distances are arranged in a specific order, and once the starting point is determined the rest fall into place. With the advent of modern computers, the task of reconstructing the protein's shape from a good diffraction pattern is trivial. The difficulty is in obtaining pure crystals because many proteins resist crystallization. Osamu Shimomura had meticulously accomplished the task of forming crystals of green fluorescent protein over two decades earlier.[11]

Getting high-resolution diffraction patterns requires a high-powered source of X-rays. Building such X-ray sources in a university setting is impractical. But it turns out that when physicists smash atoms in large super colliders called cyclotrons, a by-product is intense X-rays. These cyclotrons are loops, a half mile in diameter, which spin particles around at dizzying speeds and then smash the atoms together. The X-ray by-products can be collected by biologists without interfering with the physicists' work. There are only a few cyclotrons in the United States, so crystallography scientists wait in line to use the device to diffract their crystals.

Tsien wanted to obtain the crystal structure of green fluorescent protein because it would provide him with a three-dimensional image of the protein. He then could make assumptions about what causes the fluorophore to fluoresce and how it could be modified so as to design even better fluorescent proteins. Crystallography, however, was not a field in which Tsien had expertise. He eventually found James Remington at Oregon State University to help in the effort. Tsien lured him to collaborate by offering the opportunity to work with his newly minted eGFP. There would be no competition while they were determining the crystal structure of eGFP because the other groups were working on the crystal structure of the native jellyfish GFP.[12]

The crystal structures of eGFP and GFP, when published, would be virtually identical except for a few important differences in the fluorophore. The structures revealed a very unusual protein shape. And visualizing the structure also made it possible to answer many of the questions about how and why the proteins functioned the way they did. The natural jellyfish GFP and eGFP are beta-barreled proteins. Such a protein has a barrel-shaped structure with 11 protein strands weaving around the outer surface. A piece of the protein chain shoots directly through the middle of the barrel. The protein is 30 angstroms (Å) in diameter and 40 Å in height. To put this in size perspective, there are 10 billion angstroms in every meter, so it would take a stack of half a million green fluorescent protein molecules piled on top of one another to stand 2 millimeters high. The barrel structure is completed by two small "caps" that cover

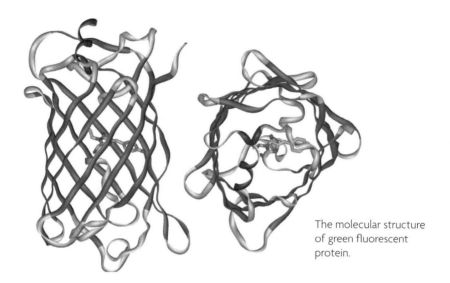

The molecular structure of green fluorescent protein.

each end. The final configuration is a compact and stable protein with a high level of symmetry. The amino acids on the outside of the barrel alternate between pointing inside the barrel and pointing outward. Positioned near the geometric center of the protein, as part of the alpha helix that passes through the center, are the three amino acids in the sequence that form the fluorophore: serine, tyrosine, and glycine.

The crystal structure of green fluorescent protein also revealed the exact arrangement of the fluorophore, one that fully confirmed the model proposed by Osamu Shimomura and refined by William Ward. The fluorophore was found to lie in a plane within the center of the barrel with several of the amino acids from the surrounding sheets pointing inward and binding to the fluorophore. That arrangement helped explain why the fluorophore, when synthesized on its own without the surrounding protein, can never fluorescence. Amino acids from the inner barrel walls coordinate the fluorophore, causing it to fluoresce. So the rest of the protein, while not necessary for the formation of the fluorophore, is essential for its function. The final piece of positive news came from the fact that both ends of the protein sequence emerged from the protein at virtually the same spot, on the top of the barrel. This finding provided hope that the sequence of GFP could be connected to or inserted into another protein's sequence without disturbing the protein's function, which would allow fluorescent tagging of proteins with fluorescent proteins.

While examining the crystal structure of green fluorescing protein, Tsien noticed an empty cavity next to the fluorophore in the center of the protein that is filled with water molecules. He suggested filling this hole with an aromatic amino acid to lower the energy level of the fluorophore electrons, changing its excitation/emission spectrum. The idea was to change a pivotal amino acid in the protein's sequence into one with an aromatic ring (only the amino acids tyrosine, tryptophan, and phenylalanine have aromatic rings). This ring might then fill this hole in the protein and stack on top of the fluorophore and reduce the fluorophore's energy levels.

This job was assigned to Andrew Cubitt, a new member of Tsien's laboratory. Cubitt had arrived in the United States in 1987 after finishing a Ph.D. in biochemistry at the University of Sheffield in England. At first he was working down the hall from Tsien. His research consisted of inserting Aequorin, the bioluminescent photoprotein from the jellyfish, into a slime mold called *Dictyostelium*, and measuring changes in calcium levels during the mold's development. While walking past Tsien's laboratory in 1994, Cubitt heard Tsien and Heim discussing the results of an experiment. "Hey, this is interesting, come check it out," Tsien said. For a brief moment, Tsien thought Heim had created a red fluorescent protein. It turned out not to be a red protein. But Cubitt was inspired by this glimpse of the fluorescent protein research and began attending Tsien's laboratory meetings; he joined the group soon thereafter. Cubitt's first task was to fill the cavity seen in the crystal structure. He created the gene for a new protein, plugged that gene into bacteria and let the bacteria grow overnight. In the morning, he examined the bacterial colonies using the "Green Monster." One of the mutants gave off a golden glow. The presence of the aromatic amino acid allowed the fluorophore to absorb lower-energy light, blue-green, and emit even-lower energy light, yellow. Yellow fluorescent protein had just been born.

With the crystal structures and this new mutant, Tsien, Remington, and Cubitt wrote up the results and submitted the paper to *Science,* but it was rejected. Tsien was dismayed and recalls:

> It went to two referees. One was a crystallographer who knew nothing about GFP (which at that time still wasn't famous enough) and said, "It's a crystallography paper but I don't know why this deserves to be in *Science.*" The second referee said this doesn't shed any light on why the jellyfish has the protein. That's still unanswered. How the hell could a crystal structure tell you why the jellyfish has the protein? Then the referee complained that we had only quoted Shimomura and Prasher and had not cited enough of the early history of GFP, including some crucial papers by [James] Morin and [Woodland] Hastings on the function of GFP. They blamed us for crediting Shimomura for discovering GFP. They said we should credit Morin and Hastings. I smelled who the referees were.

Tsien was at a loss. He tried contacting Daniel Koshland, Jr., at Berkeley, who had been an editor of *Science* from 1985 to 1994, for assistance, but he couldn't help. "Finally, on the Internet there was a discussion group already set up on fluorescent proteins and Fan Yang (George Phillips's postdoc) posted, 'Hey everybody, we solved the structure of GFP and it's going to be out in the October *Nature [Biotechnology]*,' and I took that posting on the Internet and forwarded it to *Science* and the next day it [Tsien, Remington, and Cubitt's paper] was accepted. So, it's clear that it was all politics. We hadn't changed the science . . . *Nature [Biotechnology]* and *Science* are deadly rivals. Neither one wants to be scooped by the other, so *Science* took the paper the next day."[13]

Tsien and others would go on to perform massive planned and random mutation experiments to create new and useful fluorescent proteins. This work would result in 20 different published mutants with altered and improved spectral properties. Researchers found blue, cyan, and yellow fluorescent variants with this approach. Other types of mutants were also discovered that had different sensitivities to pH and temperature, ones in which the protein formed faster at mammalian body temperature, and ones that were brighter. "The holy grail for us at that time," Cubitt remembers, "was making a red fluorescent protein because we felt that it would be much easier to work with than the blue and green ones."[14]

Today, most scientists use Tsien's fluorescent proteins, including eGFP, yellow fluorescing protein, and cyan fluorescing protein. Clontech, Inc., a southern California biotech company, paid large licensing fees to Columbia University, Woods Hole Oceanographic Institution, and the University of California, San Diego—a portion reaching Chalfie, Prasher, Tsien, and Heim. The company rapidly produced a series of user-friendly plasmids that scientists could purchase. With these improvements and the ready availability of fluorescent proteins, their use by researchers mushroomed. From the discovery of green fluorescent protein in 1962 to 1994, the year of Chalfie's breakthrough paper, there were 20 publications concerning fluorescent proteins. By 2006 that number will be over 24,000, and it continues to rise.

In 1995 Roger Tsien; another scientist in his building, Charles Zuker; along with venture capitalists founded Aurora Biosciences. Many of its technologies centered on the new fluorescent proteins. In June 1997, Aurora went public, and by March 1999, the company had a market capitalization of $1.5 billion. Many people and organizations were attracted to Aurora, including both Heim and Cubitt, who left Tsien's laboratory to work for the lucrative company. Those associated with the licensing patents for fluorescent proteins received substantial monetary rewards. One person who feels shunned is Paul Brehm, who introduced Chalfie to the glowing protein. He was not mentioned in the landmark *Science* paper, nor listed on the patent. Brehm, now a professor at Stony Brook University, recalled a different version of the 1988 meeting that Chalfie calls "murky." Brehm says he met privately with Chalfie following his seminar and they discussed problems labeling live cells. "I told him my idea of using GFP," says Brehm.[15] According to patent law, if the idea was conveyed in a private conversation, Brehm would be entitled to revenue from the lucrative green fluorescent protein patent. Disputes like this are commonplace in the idea-driven scientific world.

Although Tsien says that his research is "often dismissed as technology development, inferior to pure biology," his techniques have indisputably helped solve a wide range of longstanding biological questions. In the June 27, 1997, issue of *Science*, a news article headed "Jellyfish Proteins Light Up Cells," described the continuing effort to reengineer the heavily studied jellyfish protein. The article says: "Tsien and others are now working frantically to make mutants that glow brighter, fluoresce in more colors, or hook onto calcium ions and phosphate groups in cells and tissues."[16]

Why have fluorescent proteins become paramount research tools? GFP-like proteins are still the only proteins known that spontaneously produce visible fluorescent fluorophores after translation. All other fluorescent proteins require the addition of complex cofactors to the peptide backbone. These complex structures cannot be synthesized readily by cells and require multistep enzymatic synthetic pathways involving numerous proteins. Therefore, only GFP-like proteins offer a simple, stand-alone structure that can be planted into cells by adding only a small amount of DNA. With the cloning of the GFP's DNA sequence and the demonstration by Chalfie that the protein could be expressed in an organism other than the jellyfish, the floodgates opened. Within a few years of the Chalfie paper, clones containing the protein were in use throughout the world. The patents that Chalfie, and later Tsien, had been awarded were licensed to Clontech, which produced a variety of plasmids containing a number of permutations of the protein's DNA. Unfortunately, however, for the recombinant DNA business, plasmids are DNA and are therefore easily propagated, and as a result they are sent around informally although not always legally among researchers. The amount of DNA produced by a few milliliters of bacteria can be sent to hundreds of colleagues, who can again propagate the plasmid ad infinitum. A tiny drop of a DNA solution can even be applied to the paper of a letter and mailed to a colleague.

As of 2005, fluorescent proteins have been fused to thousands of different proteins and examined in cells. Fluorescent proteins are capable of glowing in plants, bacteria, birds, amphibians, mammals, fish, nematodes, cyclostomata (lampreys), protozoa, slime molds, and yeast, to name a few. In fact, there seems to be no form of life that cannot be coaxed to produce fluorescent proteins. The variety of their uses is staggering and ever expanding. And since patents don't restrain the activity of academic scientists, they are free to alter the DNA sequence of green fluorescent protein in any way they please. The jellyfish's green fluorescent protein is a very innocuous protein when expressed as a reporter, which allows its widespread use in a number of fields. Within a few years after Chalfie's famous paper, there were transgenic animals of all flavors glowing with fluorescent proteins. There are many proteins designed to work best in specific animals: nematodes *(Caenorhabditis elegans)*, fruit flies *(Drosophila)*, plants, fish, and mice. The use of green fluorescent protein as a scientific tool appears to be limited only by the creativity of the scientist using it. As word of its success began to appear in the scientific literature, everyone wanted to find out about the intracellular disposition of his or her favorite proteins. Soon cancer biologists, immunologists, virologists, neurobiologists, cell biologists, and even botanists were using fluorescent proteins in scientific studies.

Still, Tsien was not fully content. He wanted to push the boundaries of fluorescent proteins even further. Could fluorescent proteins monitor changes in cells? Could they provide a visual signal when calcium is increased? His dyes had done that and Ridgway and Ashley had first used Aequorin to do the same thing. His calcium dyes and Ridgway and Ashley's Aequorin technique required physically introducing foreign molecules to the living cell, often a fatal process. Tsien hoped to use green fluorescent protein to monitor calcium levels. But though green fluorescent protein's sturdiness and durability made it a popular tagging molecule, they also made it a poor candidate to monitor change. The protein's fluorescence remained stable under most conditions.

Tsien recalled a 1971 study describing the function of green fluores-

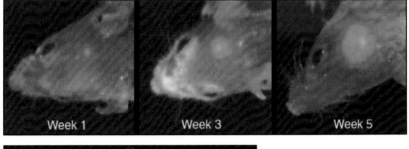

Week 1   Week 3   Week 5

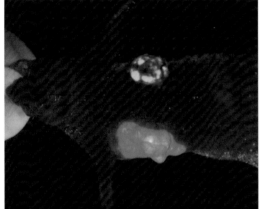

Above: A live mouse containing the gene for red fluorescent protein in brain tumor cells. The tumor grew dramatically in 5 weeks. Left: A live mouse with lung tumors labeled with red and green fluorescent proteins. Photos by AntiCancer, Inc.

cent protein in jellyfish and hydroids.[17] In these animals it converted the blue bioluminescent light into visible green light. This explained why Shimomura saw blue light from the jellyfish's purified bioluminescent protein—and not the green glow given off by the living animal. This exchange of energy is a process called bioluminescence resonance energy transfer. In order for this exchange of energy to occur, the molecule converting the light has to be extremely close to the light-producing molecule. In the case of the jellyfish, its Aequorin passes most of its energy to the accepting fluorescent protein. Tsien wondered if he could use fluorescence resonance energy transfer, or FRET, to monitor calcium levels. In this case, two fluorescent molecules share energy, rather than there being a bioluminescent molecule and a fluorescent molecule.

FRET would not be possible with only one green fluorescent protein, because the process requires two separate proteins with complementary fluorescent colors. Since Tsien had created the cyan and yellow fluorescent proteins, he wondered if he could pair them up as FRET partners. When cyan fluorescent protein is excited by violet-blue light, it gives off a cyan (bluish-green) light. But when cyan fluorescent protein and yellow fluorescent protein are close together, the same blue light should produce yellow light (as the cyan fluorescent protein transfers its energy to the yellow fluorescent protein). To test this, Tsien created a new protein by fusing cyan and yellow fluorescent proteins. When this fusion protein was inserted into cells, sure enough, blue light produced yellow fluorescence.

The next step was to devise a way to make the distance between the two fluorescent proteins change when levels of calcium in a cell changed. In this situation, a different color of light would be seen depending on the presence of calcium. To do this Tsien, co-opted two other pieces of known protein sequence. Calmodulin is a protein found in most mammalian cells that adheres to calcium. In muscle cells, when calmodulin binds calcium it then sticks to another protein (myosin light chain kinase) that regulates muscle contraction. Tsien assembled a large new protein that contained both cyan and yellow fluorescent protein, separated by both binding proteins (calmodulin and myosin light chain kinase). The idea behind this complicated new fusion protein was that when calcium levels are raised, the two binding domains come together.

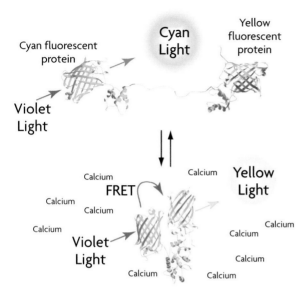

Cyan fluorescent protein

Cyan Light

Yellow fluorescent protein

Violet Light

Calcium
Calcium
FRET
Calcium
Calcium
Calcium
Violet Light
Calcium
Calcium

Calcium
Yellow Light

Calcium
Calcium
Calcium
Calcium
Calcium

An artificial protein created by fusing the DNA for two mutated fluorescent proteins (cyan fluorescent protein and yellow fluorescent protein) together with a piece of protein that binds calcium ions and another piece of protein that binds the calcium-binding piece of protein only when calcium is present. When calcium is present in cells it will change the color of its fluorescent light from a cyan to a yellow color.

After binding took place, the protein would jackknife in the middle, and the cyan and yellow fluorescent proteins would press together. When calcium is not present, the cyan and yellow fluorescent proteins are too far apart to share energy. If blue light is shined on the cell and calcium levels are low, the cell gives off only cyan light. But when calcium levels rise, blue light produces yellow fluorescence. After testing a few prototypes, Tsien found one that worked. He named his newly engineered creation cameleon, combining calcium with the word chameleon, for the color-changing lizard.[18]

Since its development, the cameleon fusion protein has been inserted into many different animals, and it allows the calcium levels in only the targeted cells to be directly monitored. It has enabled scientists to visualize the calcium waves across a beating heart. Calcium pulses can be seen in the muscles of moving flies and calcium dynamics can be witnessed in insulin-secreting cells of the pancreas. Using the ability to target a specific protein located within the cell, fluorescent protein–based sensors are providing new insights into calcium dynamics, never before observable. Now physiology can be performed in living organisms.

Not all uses of fluorescent proteins have met with widespread acceptance. In 2000, Eduardo Kac, a French artist, displayed a transgenic rabbit as an art exhibit. He commissioned Louis-Marie Houdebine and Patrick Prunet from the Institut National de la Recherche Argonomique (INRA; National Institute of Agronomic Research) to produce a rabbit that expressed green fluorescent protein. The animal was produced by injecting the cDNA for GFP into a fertilized egg and transplanting the egg into a pseudo-pregnant female rabbit (one experiencing a state resembling pregnancy that occurs after an unproductive copulation). The resulting rabbit, named Alba, contained a copy of the GFP gene randomly inserted into the mother's DNA. It glowed green when illuminated with violet or blue light. The process didn't seem to harm the animal in any way. To scientists, the mere production of a transgenic mammal such as Alba wasn't controversial. Hundreds of transgenic mouse lines have been produced by the same process. Even a transgenic monkey expressing green fluorescent protein was produced for a study.

The controversy surrounding the Alba project was that the animal was created for strictly artistic purposes, with no intent to gain scientific knowledge. The artist wrote copious position papers and a treatise on the sociopolitical ramifications of transgenic animals. Several news organizations treated the stunt as the beginning of the production of "designer" genetically modified animals. The publicity evoked a strong negative reaction from people and organizations that believe that any genetic manipulation is inherently wrong. One argument is that these genetically modified organisms could escape into the natural population and wreak havoc on the natural ecological communities. Such questions were addressed in the late 1970s, when molecular biologists were using a great deal of genetically modified bacteria in their studies. Great fear was expressed that antibiotic-resistant bacteria, engineered for molecular biology studies, would eventually end up being released into the wild and causing catastrophic disease. Governments reacted to such fears by establishing rules about the way such organisms were produced; as a result, disabling genes were often incorporated into the experimental bacteria strains to reduce their viability. Now, over 30 years later, with the nearly ubiquitous presence of recombinant bacterial strains in laboratories worldwide, no significant pathogenic or environmental impact has been seen.

The introduction of foreign genes into mammals evokes the same fears. Some believe such modified organisms may alter the natural environment and cause catastrophic environmental damage similar to that caused by the well-publicized invasions of exotic animals. The disastrous results of the introduction of the European rabbit into Australia—the rabbits bred extremely rapidly and were incredibly destructive, inflicting great damage on large tracts of productive land—highlights this problem. Yet the idea that the release of rabbits that express a single transplanted gene could harm native populations has little merit when you consider that domestication and selective breeding of animals have taken place for thousands of years. Although selective breeding does not strictly introduce foreign genes, the process selects for naturally occurring but exceedingly rare mutations. This extreme selection process has

GloFish, the first genetically modified pet available in the United States. The zebrafish *(Danio rerio)* now possesses the gene for red fluorescent protein. The dark fish on the far right near the bottom of the image has not been genetically modified and is thus barely visible. Photo by the authors.

resulted in the generation of radically different strains of the same species. Such a process, applied to the species *Canis familaris,* has produced strains as radically different as dachshunds, great danes, and chihuahuas. The spread of feral animals that are not genetically modified and of domesticated livestock that establish wild populations already produces significant environmental damage. So, in essence, this type of mass selective breeding and escape of domesticated animals has been going on for thousands of years. It is unlikely that the release of animals with single gene additions will substantially increase the already destructive environmental effects of human domestication and alteration of animals.

If the Alba rabbit raised questions about the ethics of creating designer animals, the introduction in 2002 of fluorescent protein genes into a tropical fish strain (*Danio rerio,* zebrafish) has raised greater fears. The animals, sold by Yorktown Technologies under the trade name GloFish, are meant for tropical fish collectors. The U.S. Food and Drug Administration declared that the fish are not a risk, making them the nation's first genetically altered household pets. The animals were created in Singapore, but the state of California and a number of countries forbade their sale, fearing these genetically modified animals, called by critics "Franken fish," might escape into the wild.[19] When illuminated with ultraviolet light in a dark room, these fish give off a bright fluorescent glow.

In another controversial application of green fluorescent protein, plant scientists have been developing a range of commercial plant strains that express fluorescent proteins. In some cases, the fluorescent protein is expressed in the plant along with some other gene that is inserted into the plant's genome to increase the viability or commercial value of the plant. The fluorescent protein allows rapid detection of the presence of the other gene. A whole range of enhancing genes have been devised that either make plants more hearty or more resistant to insects and other pathogens, or make the plants more nutritious or aesthetically attractive.

In 1999, the modification of the jellyfish's green fluorescent protein had produced an entire toolbox of molecular techniques for further exploration of the science of life. But one color was still missing from the palette, red, a color that Cubitt calls the "holy grail" owing to the explosive potential it holds, both as an additional FRET partner and for its unique ability to penetrate deep into cells and tissue. But no matter how researchers twisted or manipulated the jellyfish's fluorescent protein, it would not produce crimson.

CHAPTER NINE

A Rosy Dawn

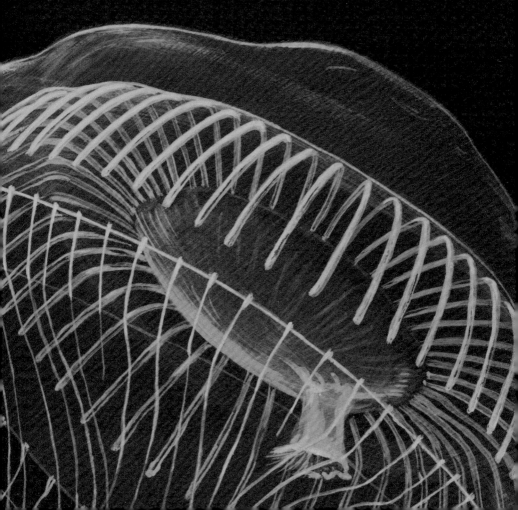

"THE DESIGN for the video game *Doom* was generated from this place," joked the molecular biologist Sergey Lukyanov as he swiftly maneuvered the stark hallways of the Shemyakin-Ovchinnikov Institute of Bioorganic Chemistry (SOIBC), a division of the Russian Academy of Sciences.[1] Located in the southwest district of Moscow, 15 kilometers from the Kremlin, the Institute was constructed in the early 1980s with funds from the Soviet Ministry of Defense. It was once a bustling hub of science, harboring more than 1,000 researchers in its helical edifice, designed to resemble the structure of DNA. While David Gruber was visiting in March 2004, the building seemed like a slumbering giant; it was only a third occupied and many of the corridor lights were turned off to conserve energy. Shortly after the collapse of the Soviet Union in December 1991, the government's support for the Institute dwindled and researchers' salaries were cut drastically. The pay—equivalent to about $50 per week—was insufficient to cover basic living costs, and many highly trained scientists fled the country to escape the bleak economic prospects.

Lukyanov, though, never considered the idea of leaving Russia. Very nationalistic, he has roots embedded deep in Moscow, where he was born and raised. "It is like the classical seed-dispersal theory," he says. "Some are designed to colonize new areas. I am not meant to stray too far from the tree." Lukyanov, a wiry chain-smoker with shortly cropped hair peppered with gray, has worked in the SOIBC for over 18 years, progressing from research assistant to graduate student to head of a 30-member laboratory. He was inducted into the Russian Academy of Sciences in 2003, at the age of 39, one of the youngest ever to be made a member of the prestigious Academy. While many scientists fled in the 1990s, Lukyanov used the government's neglect of science as an opportunity to seize new scientific freedom. "I could now set my own research agenda," he said. Quickly embracing the capitalist spirit, Lukyanov searched for creative ways to pay his bills and fund his science, including growing mushrooms and selling them at the local market.

In 1994 Lukyanov completed his Ph.D. on the development of new techniques and methods of uncovering gene function, a specialty he calls "gene hunting."[2] His work pioneered the molecular technologies that underpin subtractive cloning. These methods uncover subtle differences in how genes are transcribed by cells under different conditions. He applied these techniques while studying the small, semi-transparent flatworm *Girardia tigrina*, known for its remarkable regenerative qualities. When the worm is sliced into several pieces, each chopped segment develops into a new animal within 2 weeks. The mechanism of this regenerative feat has puzzled scientists for decades. As a molecular biologist, Lukyanov was interested in pinpointing the exact genes the worm turns on to initiate the regeneration of new body parts. To answer this question, he devised subtractive cloning, a crafty technique that subtracts the genes expressed in an intact worm from those expressed in a freshly slashed worm. The remaining genes are those specifically turned on in the regeneration process. Lukyanov's method has widespread utility, extending far beyond studies on worms. Subtraction cloning can be used to hunt down the genes involved in any cell transformation process, such as from healthy to cancerous.

Just as he was finishing his graduate work in 1994, Lukyanov was contacted by two Russian friends who had emigrated to the United States to work for Clontech, the prominent biotechnology company. During its first decade, sales of the company's tools for isolating and analyzing genes grew more than 35 percent a year. Clontech recognized the widespread utility of Lukyanov's gene-hunting methods. In 1994 it agreed to pay him about $35,000 a year in exchange for full patent rights to any techniques developed in his laboratory. Lukyanov quickly accepted the offer because in Russia the money—although paltry by American standards—would suffice to keep his small research group afloat. The arrangement also gave Lukyanov freedom to develop creatively any method related to gene hunting. This type of partnership is extremely rare in the United States because of intellectual property rules and regulations having to do with conflict of interest imposed by universities.

Over the next four years, several of Clontech's flagship products blossomed directly as a result of its partnership with Lukyanov, including PCR Select® subtraction cDNA cloning, RACE (Marathon Cloning®), and SMART® (a cDNA amplification kit). These proprietary technologies made the process of gene hunting vastly more accessible and much easier. The cookbook-style kits provided all the necessary reagents and gave step-by-step instructions, removing much of the uncertainty associated with molecular biology. Lukyanov estimates that these kits earned over $10 million for Clontech. This figure reflects the growing popularity of "kit-based" research by scientists of all disciplines who have begun to use molecular techniques in everyday research. Among the genes that researchers have isolated using Clontech products are those for breast cancer, colon cancer, cystic fibrosis, deafness, and blindness.

During his 4-year collaboration with Clontech, Lukyanov traveled to California several times. During each trip, Clontech aggressively tried to recruit Lukyanov to work permanently for the company in California. But, in 1998, once he saved enough money to purchase an apartment in his beloved Moscow, he terminated the arrangement with Clontech. While Gruber was interviewing Lukyanov in 2004 as he barreled his Lada station wagon down the chaotic streets of Moscow—aggressively changing lanes—Lukyanov said: "This is real driving. The driving in California is so boring it can put you to sleep."

Upon entering Lukyanov's sixth-floor laboratory, you leave behind the dormant atmosphere that pervades most of SOIBC and arrive in a caffeine-driven work zone. The laboratory is well equipped with state-of-the-art gene-sequencing equipment that rarely remains idle. The main office is cluttered with computers but radiates a homey atmosphere, with movies and books lining the shelves. Most of the people who work in the laboratory are Lukyanov's friends or family members, including his brother Constantine (a senior scientist in the group) and Lukyanov's first two wives. They all attended the same middle school together, School 520 in southwest Moscow. While at school, they were inspired by the same biology instructor, who led the then 14-year-old students on field trips to the White Sea, located near the Arctic Circle, to examine marine organisms.

Mikhail Matz, a prominent member of the laboratory, took the class in 1985 with Lukyanov's brother before completing his Ph.D. and postdoc work with Lukyanov. Lukyanov describes Matz as a fun-loving and adventurous man who works in the laboratory in intense short bursts, then retreats from science by playing video games, taking flying lessons, and

The scientists who first cloned the red fluorescent protein from a coral-like animal; from the left Sergey Lukyanov, Yulii Labas, Mikhail Matz, and Arkady Fradkov.

playing in a jazz band in Moscow's smoky nightclubs. Matz often wears a large silver hoop earring and a kangaroo-skin hat that he acquired in Australia.

In 1998, Matz was completing his Ph.D. work in the lab when Lukyanov called from Clontech's office in California. Lukyanov asked Matz if he would be interested in shifting his focus to cloning new types of green fluorescent protein. While working in the United States, Lukyanov had seen the popularity and profitability of GFP. Clontech had purchased the rights to Chalfie's and Tsien's patents on the use of GFP and had been successfully marketing the popular protein. Lukyanov also knew that another GFP had been found in a sea pansy *(Renilla reniformis)*, but a legal battle ensued between its discoverer and Clontech, preventing it from being introduced to the market.

"I would be absolutely delighted," replied Matz, excited to venture from landlocked Moscow in search of bioluminescent organisms.[3] Lukyanov had trained Matz to be an expert gene hunter able to identify and pluck rare genes from tiny amounts of living material. Matz's expertise was especially valuable because although many aspects of molecular biotechnology are accessible to most scientists, the isolation and cloning of very rare and novel gene transcripts from limited amounts of tissue is still an extremely difficult task. There is a method of amplifying rare sequences of DNA called polymerase chain reaction (PRC), but to use it one has to know the sequence of the gene beforehand.

Soon after receiving Lukyanov's phone call, Matz boarded a train for a two-day journey to the White Sea, a place he remembered seeing bioluminescent ctenophores, palm-sized gelatinous animals (comb jellies) without stinging cells, during his middle school field trips. Armed with a few buckets, Matz fished ctenophores from the icy water and extracted their RNA, which he later used to screen for GFP. "This was a complete failure," Matz later recalled of his trip. "It was basically a waste of time." Fluorescent proteins have yet to be found in ctenophores.

After the failed expedition, Matz and Lukyanov realized that they needed outside advice for a more directed search. Lukyanov remembered a well-rounded and insatiably curious biologist, Yulii Labas, who had helped him with embryology during his undergraduate days at

Moscow State University. Labas, born in 1933, had spent a large portion of his scientific career working under Stalin's repressive reign. During his Ph.D. studies at the Pavlov Institute of Physiology in Leningrad, he was abruptly told to discontinue his research on color vision in fish (which he refers to as his "first love") and was forced to conduct research on how to unobtrusively monitor movement.[4] As soon as he finished his Ph.D. in 1959, he moved to the Belomor Biological Station near the White Sea, "to be as far as possible from the Institute of Physiology, where Stalin's repressions of biologists could still be felt." He adds: "People who made their research careers during the time of these repressions were still in power at Pavlov's at the time." Working with ctenophores, he developed the first microelectrodes that simultaneously measured bioluminescence and cellular electrical activity.

Over the course of his career, Labas worked on various scientific subjects, ranging from electrophysiology to bioluminescence. He is known in the Russian scientific community for his unfailing presence at conferences and meetings. Described as "eclectic" and "a bit crazy" by his coworkers, Labas has a wide range of scientific interests, and his breadth of knowledge and experience makes him a fountain of ideas. He is known to approach scientific questions in a creative fashion, sometimes with his rationale not clearly evident. Labas attributes his approach to science to his father, Aleksandr Labas, a prominent painter who was hailed by the director of the Pushkin Fine Arts Museum as one of Russia's "best masters in the last century."[5]

"Yulii Labas is kind of an elderly guy going around Moscow institutions and producing lots and lots of ideas," Matz says. "Most of them sound like complete madness, but some of them are very good. So you really have to filter out what he gives you. But if you do it right, you really can get a lot out of it."

Labas was excited and flattered when Lukyanov asked him to be in-

Yulii Labas, the eccentric scientist who masterminded the search for fluorescent proteins in corals. Labas is standing in front of his portrait—painted by his father, Aleksandr Labas. Courtesy of Yulii Labas.

volved in the project. At their first meeting, Matz and Lukyanov told Labas of their efforts to search for new fluorescent proteins in other bioluminescent animals. Labas suggested a different approach, concentrating not on other bioluminescent animals but on nonbioluminescent relatives of *Aequorea* such as corals and sea anemones. Matz stared back at Labas in confusion: "What would be the point of having a fluorescent protein in a nonbioluminescent organism?" This idea seemed heretical to Matz. All green fluorescent proteins found up until this point were involved in the process of light emission—they convert the blue photons given off by the jellyfish's luciferase into a green light.[6] In *Aequorea*, GFP is present only in those cells in the animal that produce light (photocytes), and within those cells it is in tight association with the luciferase in order to perform the color conversion. Within the bioluminescence community, the function of GFP had never been questioned.

Labas, however, reasoned that since there are completely different luciferase enzymes and luciferin substrates in different bioluminescent organisms, bioluminescence had evolved independently in many different organisms. These animals had diverged evolutionarily from one another over 550 million year ago. Therefore, there seemed no reason to think that other bioluminescent organisms employed a GFP-like protein to perform the blue-to-green light conversion. In fact, no other GFP-like proteins had been found in other bioluminescent systems. Labas made another connection. On visits to the home of his friend Andrey Romanko, a local saltwater fish tank aficionado, he remembered seeing the brilliant fluorescent colors exhibited by some of the corals. Labas told Matz: "These corals are fluorescent like there is no tomorrow. These must be GFPs. 'Get out of here,' replied Matz. 'That's another crazy idea of yours.'"

Labas grabbed Matz by the sleeve and dragged him to Romanko's apartment in Novogireyevo, on the western edge of Moscow. On the outside, the building is spattered with graffiti, but inside Romanko's apartment they found a fastidiously kept and delicate coral reef ecosystem in a 100-gallon aquarium, a fantastic collection of tropical corals glowing in neon yellows, oranges, blues, and red.

Romanko first developed an interest in corals during a visit to the Moscow Aquarium. Although he never left Russia or visited a coral reef, he was attracted to the corals' "unearthly and ancient forms."[7] Trained as an engineer, Romanko has a keen sense of detail and also enjoys cultivating his elaborate orchid collection. After visiting the aquarium, he decided to expand his hobby to include corals. This hobby soon consumed Romanko and he quit his job to tend the corals full time, installing reef aquariums all around Moscow.

Matz took samples from a green anemone and several thumb-sized pieces of hard green coral from Romanko's fish tank and returned to the laboratory. None of the animals was bioluminescent. Matz was proficient at the molecular methods used to isolate novel mRNA transcripts from small numbers of cells. In this case, though, he had no idea what the sequence was of the mRNAs he was trying to isolate. Matz assumed that any new fluorescent proteins would have some similarity in sequence to the GFP from the jellyfish *Aequorea*. Matz used a variant of the PCR technique perfected in Lukyanov's laboratory. This method involved synthesizing short sequences of DNA (primers) that would bind to (complement) a region of the unknown DNA sequence. An enzyme is then added that uses the bound primer to make multiple DNA copies of the target mRNA. These copies are then inserted into a bacterial plasmid and amplified and sequenced. But how do you design primers to bind to a sequence that you do not know? The answer is you make an educated guess—actually, several educated guesses. Matz looked at the jellyfish DNA with a keen eye and found short segments within the gene that might be similar in the new proteins.

Matz and another scientist in the laboratory, Arkady Fradkov, performed the experiment on a green sea anemone, *Anemonia majano*. When the first bacterial plates were prepared, Fradkov began to examine them on an ultraviolet light box. Fradkov rubbed his eyes when he first saw most of the colonies glowing a bright green. In disbelief, he stuck his en-

Andrey Romanko in front of his fish tank, March 2004. Lukyanov's group used a sea anemone from this fish tank to find the first fluorescent protein in coral-like animals.

tire head under the dangerous ultraviolet light so he could get a closer look. To his amazement, the colonies were indeed glowing green. He continued to stare in disbelief until he burned his face and killed all the fluorescent colonies. "This was the happiest day of my scientific career so far," he recalls.[8] Since he had killed all the cells, Fradkov had to reperform the experiment before proudly showing the results to Lukyanov and Matz. Another belief about fluorescent proteins had been proved false. These experiments toppled the notion that fluorescent proteins were found only in bioluminescent animals.

A few months later, Romanko called Matz in the laboratory: "Look, I just received from Vietnam a red fluorescent coralli-morph; you should come, now, because this rock is for somebody who will pick it up at the end of the day. There are several little things sitting on it; you can steal one." Matz ran over and picked off a few pieces and took them back to his laboratory and generated a set of bacterial plates containing the corallimorph genes. Lukyanov examined the bacterial plates the following day. After he screened thousands of nonfluorescent colonies, something caught his eye: a single colony glowing a bright ruby red. "It was a small colony, but so beautiful, so good, like a small sun," he recalls. From this red corallimorph found in the fish tank, a species of *Discosoma*, the first red fluorescent protein had been cloned. Today it is widely used under the name dsRED. Matz's red fluorescent protein emitted light that peaked at 583 nanometers, decidedly in the red portion of the spectrum.

When they sequenced their new proteins, they discovered the distinctive features of a beta barrel, the same structure as that of GFP from the jellyfish, but their amino acid sequences were very different. When the paper describing their findings was published, Roger Tsien wrote an introductory piece for it, appropriately entitled "Rosy Dawn for Fluorescent Proteins."[9]

Tsien would later say in an interview: "There were an amazing bunch of aspects about that paper. First, they said, you don't have to be bioluminescent to express a fluorescent protein. Up until then, we had been rigidly told that fluorescence always was with bioluminescence. That was wrong."[10]

The longer wavelengths of light used to excite the red fluorescent protein made it more attractive for studies in which the protein was expressed deep within a tissue. Although red light has less energy than green, its longer wavelength penetrates deeper into biological material. For instance, infrared light generated from our bodies can be imaged through building walls. This is the basis for the heat-sensing imaging technology used by law enforcement officials and the military. With the red protein scientists now had the ability to unobtrusively see cell functioning inside living organisms, particularly the brain. This technique would provide a better picture of how healthy and diseased cells operate.

Although at the time it appeared that the detection of fluorescence in nonbioluminescent animals in 1999 was a novel finding, in fact there had been several reports over the years about the phenomenon that went largely unnoticed. As early as 1927 a scientific report appeared in *Nature* entitled "Fluorescence of Sea Anemones."[11] There, the British researcher C. E. S. Phillips reported that a species of anemone collected from Torbay, on England's southwest coast, fluoresced under ultraviolet light. Phillips did not publish any follow-up papers. In the 1940s, Siro Kawaguti, a Japanese marine biologist studying coral pigments at the Palao Tropical Biological Station in Palau, Micronesia, wrote that green was the most common fluorescent pigment of corals.[12] There were even public exhibitions dedicated to coral fluorescence. In the late 1950s, René Catala, the director of the Noumea Aquarium in the Pacific archipelago of New Caledonia, established a permanent exhibit of fluorescent corals in his aquarium. He then took the exhibition to Europe and in 1964 published a book with pictures of fluorescent corals entitled *Carnival under the Sea*. Louis Fage, a member of the French Academy of Sciences, wrote in the book's preface:

In the darkness of an evening, turning the beam of an ultra-violet lamp upon the deep-sea corals, Catala suddenly reveals an astonishing spectacle. As if touched by the wand of a fairy, all the polyps change colors. The varied and beautiful hues adorning them by daylight vanish, and are replaced by a fairyland of glittering gems which dazzle the observer. Rubies, emeralds, and topaz glow here and there at the tips of the tentacles, around the mouths, along the bodies of the flowering madrepores as their outlines become lost in the deep shadows of the tanks. On discovering such beauties, visitors throng, and scientists take notice, everyone wants to observe and study.[13]

Yet scientists still did not take notice. Then in 1995 an article appeared by Charles Mazel called "Spectral Measurements of Fluorescence Emission in Caribbean Cnidarians." Mazel described corals that "contain substances in the host tissue that fluoresce when stimulated by ultraviolet and/or visible light."[14] Mazel captured images of coral at night by using ultraviolet filtered lights and strobes. He found the images so compelling that in the 1990s he founded NightSea, Inc., a company that sells the lamps, filters, and camera attachments that allow scuba divers to take their own underwater photos of fluorescing sea life.

In 1995 a coral biologist at the University of Sydney came closer than anyone to isolating fluorescent proteins in corals. In a paper titled "Isolation and Partial Characterization of the Pink and Blue Pigments of Pocilloporid and Acroporid Corals," Sophie Dove and her coauthors described a set of proteins they called pocilloporins.[15] But they did not notice the similarities between their pocilloporins and green fluorescent protein.

It was the iconoclast Russian Yulii Labas who first made intellectual leap from jellyfish to corals. Labas's background in evolutionary biology and biochemistry, and his boundless curiosity, led him to

A copepod, a tiny crustacean, exhibiting green fluorescence as a result of the expression of a fluorescent protein. This was the first fluorescent protein found outside of a coral, jellyfish, or sea anemone. Photo by Sergey A. Lukyanov/Evrogen.

suspect fluorescent proteins might also be present in nonbioluminescent animals. Yet Labas's idea would have gone untested without the advanced gene-hunting expertise and motivations of Lukyanov's group. The process of finding and isolating a gene, especially when the sequence is unknown, requires expertise normally found outside the realm of most traditional biologists. It was a rare combination of gene-hunting skills and evolutionary knowledge that brought the Russian group to a private coral aquarium on the outskirts of Moscow.

Then in 2004 the Russians shocked the scientific community once again when they discovered fluorescent proteins in three species of copepods. These small shrimp-like animals are closely related to the bioluminescent sea firefly *Cypridina*, studied by Shimomura and Harvey. Copepods are essential components of marine communities, consuming the miniature phytoplankton cells, including dinoflagellates, and passing the sun's energy up the food chain. This was the first time a fluorescent protein had been discovered outside the phylum Cnidaria, which contains corals, jellyfish, and sea anemones.[16]

CHAPTER TEN

Shimmering Reefs

O N MAY 30, 2003, Disney/Pixar Films released *Finding Nemo*, an epic saga of a fretful clownfish in search of his son, Nemo, who had suddenly been scooped out of the Great Barrier Reef and plopped into a dentist's fish tank in Sydney. In the animated movie's first weekend, it grossed over $70 million and soon was the twelfth highest-grossing film ever made in the United States.[1] *Finding Nemo* drew unprecedented attention to coral reefs and their peculiar inhabitants. According to an Australian government report, "*Finding Nemo* boosted public interest in coral reefs, raised awareness about coral conservation and provided incentives for the industry to address the impacts of the trade in marine aquarium organisms."[2] The film also elicited a few unexpected responses. Some children tried to "liberate" their fish by flushing them down the drain. RotoRooter, a plumbing company, received over 90 calls by bewildered parents and began a "Don't Flush Nemo" campaign, educating children about the slim survival chances of a flushed fish.[3]

Finding Nemo contains many correct biological descriptions of reef animals and complex ocean circulation patterns, such as the East Australian

Current. Pixar consulted several marine scientists and biomechanics experts, who presented over 20 lectures to Pixar employees on a diverse range of topics such as fish behavior and locomotion. A few inaccuracies sifted through the cracks, such as cold-water kelp living on a warm reef, but most of the details are accurate. For instance, the female anglerfish (discussed in Chapter 1) is shown in the movie with a bioluminescent lure extending from her dorsal fin with a parasitic male clamped onto her body, just above the anal fin.

Clownfish, as the movie portrays, form an unusual partnership with certain sea anemones. Like jellyfish and corals, anemones are armed with pressure-activated harpoon-like stinging devices called nematocysts. But clownfish mysteriously weave unharmed among the stinging tentacles. It is suspected that juvenile clownfish develop immunity after repeatedly being stung, or that a mucous coating on their bodies shields them from the anemones' stings. Whatever the mechanism, clownfish are protective of their anemone hosts and aggressively challenge anemone-consuming fish.

What was not depicted in *Finding Nemo* is that anemones are brimming with several colors of fluorescent proteins. A far-red (611 nanometers) fluorescing protein was first reported from the bubble-tipped anemone, *Entacmaea quadricolor,* a few months before *Finding Nemo* was released. This protein attracted keen interest among the neuroscience community, because far-red light is capable of penetrating deeply into cells and tissues. Like the Moscow fish tank coral containing red fluorescent protein, the bubble-tipped anemone is a common inhabitant of private aquariums. Jörg Wiedenmann, a research assistant in the Department of Zoology and Endocrinology at the University of Ulm, Germany, received the anemone as a gift from the Ulm Association of Seawater Aquarists. Two years before that, Wiedenmann had completed a

A clownfish *(Amphiprion clarkii)* living among parts of a bubble-tip anemone *(Entacmaea quadricolor)*, the animal where a far-red (611 nanometers) fluorescent protein was discovered. The photo was taken by Roberto Sozzani at Sangeang Island, Indonesia.

Ph.D. at the University of Ulm, where he identified fluorescent proteins in the snakelocks anemone, *Anemonia sulcata,* native to the Atlantic Ocean and Mediterranean Sea. In his 2002 paper, Wiedenmann described his newfound far-red protein and compared it to the red fluorescent protein from the Moscow fish tank.[4]

By 2002, both green and red fluorescent proteins had become standard tools of biomedical research, although certain limitations still ham-

A fluorescent anemone, Lizard Island, Great Barrier Reef, Australia. Photo by the authors.

pered their application. For example, green fluorescent protein is difficult to view deep inside tissue. Red fluorescent protein can be viewed far beneath the skin, but it has its own set of problems. The red fluorescent protein found in Moscow tends to clump together inside cells, form slowly, and go through a green stage before turning red. Tsien's group poked and prodded the red protein found in Russia and 33 mutations later produced a nonclumping red fluorescent protein. But that new hybrid still produces only weak fluorescence.[5]

Wiedenmann's far-red protein from the anemone, unfortunately, also has a critical characteristic that hampers its medical applications. It performs poorly, if at all, at the temperature of the human body, 98.6°F. A sturdy red fluorescent protein would offer many advantages, but where could this be found? One fact was becoming increasingly evident; animals from coral reefs are hotspots for fluorescent proteins.

Fluorescent coral (*Zoanthus* species), Lizard Island, Great Barrier Reef, Australia. Photo by the authors.

And they could possibly contain the elusive "perfect" red fluorescent protein.

That same year, 2002, the authors headed to the Indo-Pacific to search for and catalogue fluorescent proteins. If a few aquarium corals and anemones had provided several new fluorescent proteins, we wondered what surprises the world's most diverse coral reef ecosystem would hold.

Lizard Island on Australia's Great Barrier Reef offered an ideal study location. Our primary goal was to inventory the fluorescence of corals and other reef creatures. One complication with the inventory is that fluorescence is best examined in a dark environment, without the interference of sunlight's broad spectrum. It seemed that diving at night was the most feasible approach to nondisruptively scan and inventory reef creatures for fluorescence. We therefore custom designed back-mounted, high-intensity underwater lighting systems and cameras to capture fluorescence. During the day, we mapped the destination using light sticks attached to buoys. Night after night, we pointed small motor boats into the darkness. We then carefully maneuvered

Fluorescent coral *(Acropora latistella)*, Lizard Island, Great Barrier Reef, Australia. Photo by the authors.

through curving grooves in the shallow reef formation until we reached a select location. There, we descended coral walls using only a single, focused beam of light. On each dive we used different colored lights and filter sets—each designed to seek out a differently colored fluorescent protein. While we were underwater, one of us managed the light source, while the other peered through the fluorescent video camera. When a glow was spotted, a thumbnail-sized sample was collected and analyzed.

Within only one month, we encountered and catalogued over a hundred different species of corals expressing yellow, green, orange, and red fluorescent proteins. The corals exhibited a wide range of fluorescent patterns. Some contained intricate floral patterns and harbored several different fluorescent proteins. Sometimes we encountered corals of the exact same species living together, one individual glowing, the other exhibiting no fluorescence. At the time this book was written, we were able to clone over 25 new fluorescent proteins and are continuing the hunt. We are still seeking the elusive "perfect" red fluorescent protein. Nevertheless, the abundance and variety of fluorescent proteins raised a perplexing question: Why do fluorescent proteins concentrate on tropical reefs?

On a thriving coral reef, almost every piece of underwater real estate is occupied by a variety of marine citizens that at first glance live in peaceful coexistence. But a closer inspection reveals thousands of sustained and merciless battles over scarce food and space. This intense competition is a driving force for the staggering biodiversity of coral reefs. Defense mechanisms and cooperative agreements are essential for survival in this seemingly idyllic setting. Unassuming cone snails slink about, armed with a syringe containing over 30 types of the most potent neurotoxins. Each night, at the onset of dusk, nocturnal creatures try to uproot coral animals from their calcium carbonate homes, while the cor-

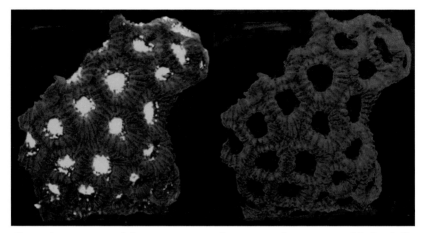

Above: Fluorescent corals (*Goniastrea* species). Opposite: Fluorescent coral *(Lobophyllia hemprichii)*. Lizard Island, Great Barrier Reef, Australia. Photos by the authors.

als release a mucous coating to protect themselves. Although coral reefs occupy less than half of one percent of the ocean bottom, they are teeming with life, the most diverse of all marine ecosystems. Estimates of the number of species found on reefs range from 600,000 to more than 9 million, which places them just behind tropical rain forests as the most diverse ecosystem on earth.[6]

Corals are an eccentric breed of creatures that were historically mistaken for plants. In 1744 Jean André Peyssonnel discovered that the "flowers" of corals were actually tiny anemone-like animals. In an unpublished manuscript he wrote: "I saw the coral flowering in vases full of seawater, and I observed that what we believed to be the flower of this so-called plant was . . . similar to a small nettle or octopus."[7]

Each of Peyssonnel's "nettles" is a single clonal polyp, an animal rang-

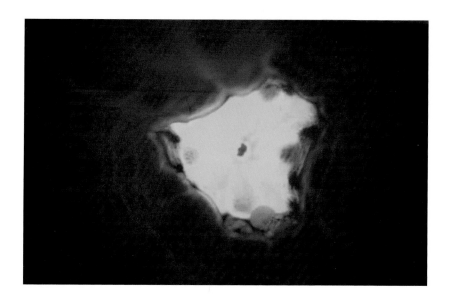

ing in size from a pinhead to a bagel. What separates corals from their close relatives sea anemones and jellyfish is that corals secrete a calcium carbonate skeleton. Scleractinian corals are the stony, reef-building corals and usually have six tentacles per polyp, while soft corals, or octocorals (eight tentacles), are more pliable but have spiked spicules beneath their endodermis, or "skin," which make them unpalatable to predators. Corals secrete calcium carbonate by methodically extracting calcium and bicarbonate ions, which are abundant in seawater. Depending on the species, coral colonies form massive boulders, branching finger-like structures, or solitary polyps. Large boulder corals grow rather slowly, only a few millimeters each year, while some branching corals accrete at more than a centimeter a year, about the same rate as fingernails grow. Each polyp divides and produces additional polyps, culminating in a network of neighboring interconnected colonies. The oldest known boulder corals are found in the Indian Ocean and originated as a single colonizing polyp over 5,000 years ago.

Most reef-building corals exist in an intimate, complex, and still largely mysterious symbiotic relationship with a bizarre class of protists called zooxanthellae. These are a type of dinoflagellate, single-celled algae that are also responsible for producing the majority of bioluminescence in the ocean and such unwelcome events as toxic red tides and paralytic shellfish poisoning. One fall morning in 1947, for example, the

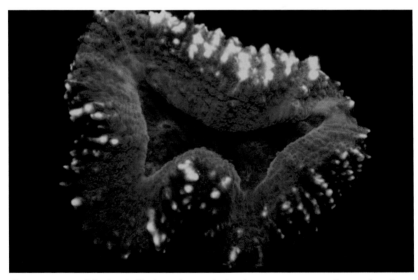

Fluorescent coral *(Lobophyllia hataii)*, Lizard Island, Great Barrier Reef, Australia. Photo by the authors.

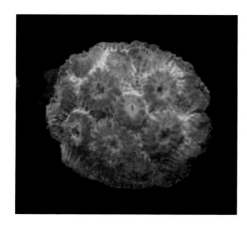

Fluorescent coral. *(Cyphastrea microphthalma)*. Photo by the authors.

community of Venice, Florida, awoke to find the sea turned into what looked like a brownish-red soup.[8] Thousands of dead fish suddenly appeared along the beaches and a stinging gas filled the air, causing respiratory troubles. At first residents blamed a chemical spill, but scientists later discovered that a bloom of a single dinoflagellate species, *Karenia brevis*, was to blame.

Only about 3 percent of known dinoflagellate species are harmful, but those that are produce some of the most vicious neurotoxins. Saxitoxin, for example, produced by the dinoflagellates *Protogonyaulax catenella* and *Gessnerium monilatum*, is 1,000 times more toxic than the potent nerve gas Sarin, used in the 1995 Tokyo subway attack. Because of its lethal character, saxitoxin is listed as a Schedule 1 toxic chemical, a designation reserved for those chemicals that pose the highest risk and have limited commercial use, according to the Convention on the Prohibition of the Development, Production, Stockpiling, and Use of Chemical Weapons and Their Destruction.[9] The agreement, also known as the

Chemical Weapons Convention, was ratified by the United States in 1997.

Zooxanthellae are nontoxic dinoflagellates that live within the endodermis of reef-building corals. Individual coral cells engulf the zooxanthellae cells (each about 1/100 of a millimeter in size) and maintain the algae in a small membrane bubble inside the coral's cell. Reef-building corals are densely packed with zooxanthellae, which give many corals a brownish color. This cohabitation—symbiosis—is advantageous to both the coral and the algae. Most important, the zooxanthellae supply the coral with food, such as glucose, glycerol, and amino acids, manufactured by the algae during photosynthesis. These substances provide the coral with its primary energy source for survival. In some instances, zooxanthellae exude more than 100 percent of the food that is necessary for coral survival. For this reason, corals have become highly dependent on their houseguests while living in nutrient-depleted azure tropical waters. In return, the coral provides the zooxanthellae with a safe, protected, and sunlit home. This mutual exchange, driven by the algae's photosynthesis, is the key to the prodigious biological productivity and limestone-secreting capacity of reef-building corals.

Given the importance to corals of photosynthesis by their algae guests, could this account for why fluorescent proteins are present in such large quantities in corals? The answer remains unclear, but one theory put forward is that fluorescent proteins enhance photosynthesis for corals that live deep in the water, where there is little available light.[10] Water is very efficient at absorbing light, especially the specific wavelengths of light necessary for photosynthesis. Below about 100 meters, there is sparse light to drive photosynthesis. But high-energy light, such

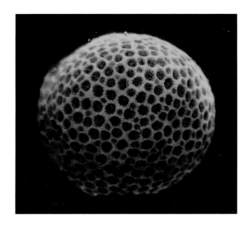

Porities sp. Lizard Island, Great Barrier Reef, Australia. Photo by the authors.

as ultraviolet light, although it does not drive photosynthesis, does penetrate deep into the water column. This theory of fluorescent protein function hinges on the idea that fluorescent proteins absorb the deeper-penetrating high-energy light and then convert it into green light. Green light can then be utilized by the algae for photosynthesis. The problem with this idea is that it does not withstand scientific scrutiny. There do not appear to be sufficient fluorescent proteins in most coral cells to transfer enough light to enhance photosynthesis.

Another idea is that the fluorescent proteins protect the coral when the sunlight is too strong; in other words, they prevent the coral from getting burned by overexposure to the dangerous rays of the sun.[11] Each day, photosynthetic machinery, including the chlorophylls of the symbiotic algae in corals, is damaged by the intense sunlight. When this occurs, the light-harvesting centers of the algae become "photo-inhibited," which reduces their ability to produce energy. If a coral had enough fluorescent proteins, the theory goes, the proteins could be mobilized to absorb light, which would reduce the amount of light striking the algae's photosynthetic centers, preventing them from becoming overwhelmed. In this scenario, the fluorescent proteins could act to protect the coral from sunlight. The jury is still out on this hypothesis.

Another theory is that fluorescent proteins protect coral from a dangerous by-product of photosynthesis. At times of intense sunlight, the chlorophylls produce too many high-energy electrons. With nowhere to go, these electrons produce free radicals, which can then react with other cell components and cause damage. Other photosynthetic organisms have evolved a number of mechanisms to protect themselves from the free-radical by-products of photosynthesis, but corals are not photosynthetic organisms and do not have such defenses. The walls of the algae and the compartment in the coral that contains them are permeable to allow the carbohydrate food from the algae to be taken up by the coral and the nutrients from the coral to be passed to the algae. This arrangement, however, also allows the dangerous free radicals to attack the coral. If the algae were planktonic, like many dinoflagellates, the free radicals would diffuse into the ocean rather than the coral. But zooxanthellae living inside a coral do not have this option, and their toxic by-products diffuse into the coral tissue. It is possible that the coral produces fluorescent proteins to absorb free radicals produced by the algae.

The relationship between zooxanthellae and coral has suddenly become a pressing scientific concern. Several years ago, coral biologists around the world began to notice that large swaths of coral were suddenly changing from a rich brownish color to a bright color, then a ghostly white. Scientists in the Caribbean, the Indian Ocean, the Great Barrier Reef, and the Red Sea all began finding the same mysterious condition. About a quarter of the corals that turned white would die within a few weeks. This came to be known as coral bleaching. Bleaching events occur worldwide, within a few weeks of each other, and always during the summer months. In 1998 and 2002 about half of the world's reefs underwent bleaching.[12] The fact that bleaching events occurred simultaneously in reefs located thousands of miles apart rules out culprits such as agricultural runoff and nutrient enrichment. Such a widespread event suggested a global factor. Since unusually hot spikes in the summer temperature accompany bleaching events, it was realized that a slight increase of only 1 or 2 degrees in temperature was sufficient to trigger coral bleaching. Global warming, pollution, and events such as El Niño are thought to have a large impact on the health of zooxanthellae and thus coral reefs worldwide.[13]

Coral bleaching has become the number-one focus of coral ecologists because it is the major assault on the world's coral reefs. Damage by tour boats and coral collectors is dwarfed by comparison. Scleractinian reef-building corals are at the heart of the reef's structure and ecosystem, housing thousands of different organisms. Without the shelter and structure provided by the corals, a reef's diversity plummets. Scientists fear that without a reduction in the amount of bleaching 60 percent of coral reefs may disappear by 2030.[14] Skeptics within the field point out that warming events have happened in the past and corals have survived them. But previous climatic changes occurred over thousands to millions of years, unlike the rapid warming experienced over the last century.

Coral bleaching results when the host coral loses the symbiotic zooxanthellae algae from its tissues. Either the coral or the zooxanthellae decide that their cohabitation is no longer mutually beneficial and the algae leave—or they are expelled. Although some corals can withstand the loss of their symbionts, most cannot and eventually starve to death. Several lines of evidence seem to indicate that the levels of free radicals increase in the corals during elevations in ocean temperatures that induce bleaching. Although free radicals form in the algae normally during photosynthesis, excessive levels spread into the coral during extended bouts of high temperature. Possibly the coral, recognizing the increased danger of excessive free radicals, opts to eject the algae rather than suffer potential damage from the algae toxins. Certainly much of the process of bleaching remains a mystery, but the discovery of fluorescent proteins in corals will probably play a crucial role in scientists' attempts to understand the physiology of coral bleaching.

The discovery of a wealth of fluorescent proteins in corals highlights the undiscovered bio-commodities in reef organisms. Although terrestrial organisms exhibit great species diversity, marine organisms have representatives of every phylum, plus several phyla and thousands of species found nowhere else. For example, tunicates, bryozoans, sponges, and echinoderms are absent from terrestrial ecosystems. The diversity of reef organisms provides scientists with a smorgasbord of life forms in which to search for elegant molecular tools that can be used to solve important biological problems. The highly competitive nature of reef life, and millions of years of evolution, have created astounding diversity in life and in modes of survival. Many reef animals are firmly attached and cannot escape environmental perturbations, predators, or other stressors. They engage in many forms of chemical warfare, using bioactive compounds to deter predation, fight disease, and battle competing organisms. Some animals also use toxins to catch their prey. These compounds are either produced by the organism or obtained from microorganisms living within their tissues. Because of their unique structures and biological activities, defensive compounds often yield life-saving medicines and industrial products.

Scientists and bioprospectors have only scratched the surface of the vast molecular and chemical library contained in the organisms that inhabit the reef environment. Intriguing hints of new compounds with enormous medical and commercial potential await further investigation. A striking example is a genus of snails, *Conus*, consisting of over 700 known species. These animals live in beautifully adorned cone-shaped shells ranging in size from less than an inch to over 3 inches. Their sluggishness masks a ferocious attitude. Different members of the genus hunt fish, other mollusks, or small invertebrates. The cone snail stalks its prey and then, with lightning speed, deploys a harpoon containing a powerful neurotoxin. The harpoon lodges in the victim and injects a potent cocktail of neuroactive agents. These agents include blockers of many mammalian proteins, including sodium, calcium, and potassium ion channels. The toxin can kill humans in a matter of minutes, but scientists have begun to harness components of the toxin to treat ailments such as heart arrhythmias, epilepsy, and severe pain. Pain treatments derived from *Conus* species may some day replace morphine and codeine, because the former are far more potent and less addictive.

Corals and sponges have also been examined for their anticancer and antimicrobial secretions. Several such agents are currently in clinical use or trials. Compounds extracted from sponges have been used as antiviral drugs to treat HIV and herpes infections. Sea fans are used to make products to relieve sunburn, and the coral *Pseudopterogorgia elisabethae*, known for its anti-inflammatory properties, is used in facial cleansers and creams. Modern molecular and biochemical purification techniques promise to make many more of these agents available to treat human diseases.

Unfortunately, 58 percent of the world's reefs are threatened by human activities.[15] In addition to global warming, intensive farming, deforestation, and development are introducing large quantities of sediment, nutrients, and other pollutants into coastal waters, causing widespread degradation of coral reefs. Coral reefs are often fished intensively; in regions of the Indian and Pacific Oceans, destructive fishing with dynamite and poisons has devastated reef habitats.[16] In Roger Tsien's piece introducing Matz's article first describing fluorescent proteins in corals, he wrote: "Coral reefs are among the most beautiful and species-rich habitats on earth but are also among the most threatened by climate changes, pollution, and shortsighted overexploitation. Although more reasons for reef preservation are hardly needed, the beautiful and useful FPs from fluorescent corals provide yet another concrete example of the practical value of biodiversity."[17]

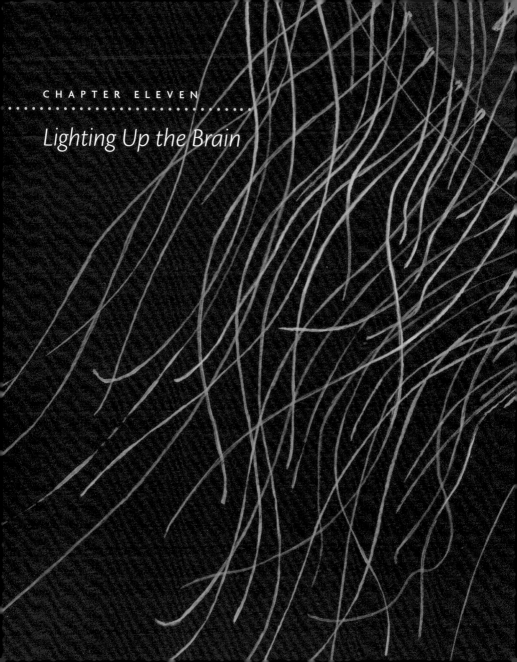

CHAPTER ELEVEN

Lighting Up the Brain

HE QUEST for new fluorescent proteins that began as a search solely in bioluminescent animals expanded to fish tanks, and then to sun-drenched coral reefs, and then to encompass almost any creature lurking in the coastal waters of the oceans. Now there appear to be no bound-aries to where researchers are looking for new and unique fluorescent proteins—even in the deepest recesses of the ocean's bottom. The scien-tific applications of fluorescent proteins have also blossomed. Only a few years after Chalfie's 1994 insertion of the jellyfish's fluorescent protein into another animal, the illuminating fluorescent molecules began pierc-ing the walls of the most intangible, mystifying, and perplexing organ: the brain.

Hidden inside the human brain, a three-pound fatty organ with the consistency of Jell-O, is a structurally complex electrical and chemi-cal universe, capable of a processing power beyond the abilities of the most sophisticated computers. Our understanding of the brain's link to thought, logic, memory, and emotion dates as far back as 400 B.C.E. when Hippocrates proclaimed the brain to be the seat of intelligence: "With the operations of understanding . . . the brain is the cause."[1] Aristotle later disagreed, believing that the heart encapsulates knowledge and the brain is merely a mechanism for cooling blood. He further reasoned that humans are more rational than other animals because they have a proportionally larger brain to cool their hot-bloodedness.[2]

Over the next two millennia, the brain continued to perplex people because its featureless appearance made it hard to believe it had a sophisticated function. As late as 1653, Henry More, a prominent Cambridge University Platonist, wrote that one can discern no more from the "lexe pithe or marrow in man's head . . . than we can discern in a Cake of Sewet or a Bowl of Curds."[3] But once a technique was developed that allowed the brain's microscopic structure to be revealed, the brain's complexity became dramatically evident. It does, indeed, have more function and sophistication than a bowl of curds.

That turning point in exploration of the brain came in 1887, when Santiago Felipe Ramón y Cajal, a 35-year-old anatomy professor at the University of Valencia, Spain, traveled northwest toward Madrid to dine at the house of a friend and learn the latest techniques for staining brain cells. His friend, Luis Simarro, a neuropsychiatrist and political activist, had recently returned from a self-imposed exile in Paris. He had left a few years earlier because Spanish authorities told him he could no longer examine the brains of dead patients at the asylum he ran.[4] Simarro brought back from France brain tissue samples stained with a cutting-edge silver chromate technique. Although the method had been invented 14 years earlier by Camillo Golgi, "Extraordinary Professor of Histology" at the University of Pavia, Italy, this was the first time Cajal had seen such detailed images of brain structures.

The Golgi stain is unique in its selectivity, mysteriously highlighting only about 1 percent of neurons in each particular tissue section. Since

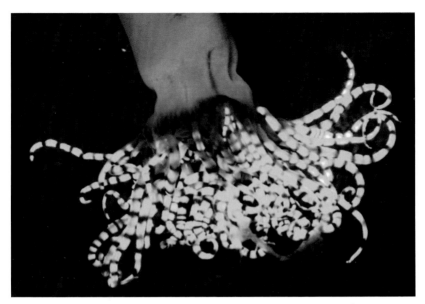

A deep-sea fluorescent anemone, Gulf of Mexico. Photo by Charles Mazel.

the brain is packed with a sinuous convolution of cells, when all cells are stained the tissue appears as an unremarkable solid black mass. With the Golgi method, only a few isolated neurons become black against a pale yellow background. "All was sharp as a sketch with Chinese ink," Cajal wrote.[5] Highlighting only a few cells clarifies their dendritic patterns and connections, unraveling the massive neuronal tangle. "It was there, in the house of Dr. Simarro," Cajal wrote, "that for the first time I had an opportunity to admire . . . those famous sections of the brain impregnated by the silver method of the Savant of Pavia."[6] Cajal was born in 1852 in the impoverished Spanish country town of Petilla de Aragon, and as an

8-year-old boy remembers having an "irresistible mania for scribbling on paper, drawing ornaments in books, daubing on walls, gates, doors, and recently painted facades, all sorts of designs, warlike scenes, and incidents of the bull ring."[7] A poor student, Cajal was also a rebellious child who exhibited a violent temper, once venting his aggression by blowing up a neighbor's gate with a homemade cannon. Cajal's lack of scholarly interest and his delinquency distressed his father, a surgeon, and at the age of 14 he was apprenticed to a barber in hopes he would learn a trade. This experience indirectly proved useful in Cajal's forays into the brain. In the barbershop, he developed an uncanny agility with a razor, a skill that would later aid in precisely slicing brain tissue sections.

Cajal's interest in anatomy first sparked when he was 16 years old. His father took him on a moonlight jaunt during which they climbed the walls of a deserted cemetery in Ayerbe in search of bones to study. There they found various skeletal remains. Cajal wrote: "In the pallid gleam of the luminary of the night, those skulls half covered with fine gravel, and with irreverent thistles and nettles clambering over them, seemed to me something like the hulk of a ship cast up on the shore."[8] They brought the bones back home, where Cajal began to sketch them. Thus began his career in anatomy.

In 1890, after two years of refining and perfecting Golgi's staining technique, Cajal published 14 original papers on the structure of the nervous system. "Realizing that I had discovered a rich field, I proceeded to take advantage of it, dedicating myself to work, no longer merely with earnestness, but with a fury," he wrote. "A fever for publication devoured me."[9] Cajal obsessively stained every type of brain tissue he could obtain: those of mice, cats, rabbits, guinea pigs, dogs, cows, sparrows, chickens, lampreys, and humans. For the first time he produced drawings of the detailed structures of the brain, providing a glimpse of its breath-taking complexity and beauty.

In the late 1880s, the field of neuroscience, still in its infancy, had largely accepted the doctrine that the nervous system was constructed of a network of continuous elements, like an elaborate structure of fused glass tubing. A fused system appeared to be the only plausible explanation of how messages could be rapidly and accurately transmitted through the nervous system. When Cajal peered through his microscope, however, he began to notice a recurrent feature: isolated nerve cells with distinct treelike appendages. These observations were incompatible with the accepted doctrine of a single interconnected structure. "As new facts appeared in my preparations, ideas boiled up and jostled each other in my mind,"[10] Cajal wrote. In his drawings Cajal revealed that nerve cells are separate units, each ending near dendrites and other cell bodies. They did not appear fused, as was previously thought, but were organized more like a bucket brigade. By applying Golgi's staining method, Cajal produced and published hundreds of graphic images that soon persuaded the vast majority of the scientific community to reject the fused theory of the brain in favor of a model of the brain as a collection of many individual neurons. The brain could be compared to a sweater made of millions of individual threads, rather than one long strand.

Oddly, Camillo Golgi, whom Cajal considered his hero, was one of the few unwavering dissenters. On December 10, 1906, Cajal and Golgi shared the

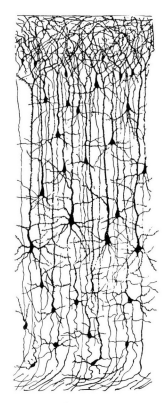

Santiago Felipe Ramón y Cajal's ink drawing of neurons in the human cerebral cortex. Copyright © Heirs of Ramón y Cajal.

Nobel Prize in Physiology or Medicine. Cajal and Golgi met in person for the first time at the award ceremony, and Cajal's image of Golgi as a great scientist was shattered when Golgi used his Nobel lecture to try, one last time, to bolster the toppled fused theory. "While I admire the brilliancy of the doctrine which is a worthy product of the high intellect of my illustrious Spanish colleague, I cannot agree with him," Golgi said. He further mocked Cajal's neuron theory by saying: "The majority of physiologists, anatomists and pathologists still support the neuron theory, and no clinician could think himself sufficiently up to date if he did not accept its ideas like articles of faith."[11] Most in the audience were dumbfounded, since this statement belied decades of research based on Golgi's own technique. Some 40 years later, after both Golgi and Cajal had died, electron microscopes revealed that tiny spaces do exist between neurons—spaces called synapses—providing unequivocal evidence that Cajal was indeed correct. Today his hand-drawn sketches of the brain are still frequently reproduced in neuroscience textbooks.

Cajal's impact on neuroscience illustrates how enhancing our ability to visualize the brain transformed our understanding. Cajal used a deceivingly simple technique to advance dramatically what was then known about the brain. The Golgi technique revealed the brain's hidden labyrinthal structures, capable of harboring consciousness. Throughout history, technological breakthroughs that made the invisible visible have played pivotal roles in science, allowing long-standing roadblocks to be overcome and paving the way for great leaps forward.

The images Cajal penned from dead tissues are remarkably similar to those seen when living brain tissues are highlighted with fluorescent proteins. But fluorescent proteins offer the benefit of tagging selective brain cells, rather than the random and haphazard staining that

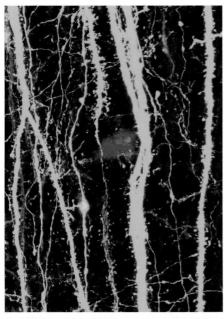

Left: A fluorescent micrograph from a living brain showing individual nerve cell dendrites (green) surrounding an Alzheimer's disease–like deposition of amyloid (red). The image was taken through a thinned region of the skull of an anesthetized mouse with a multiphoton laser scanning fluorescent microscope. Each green nerve cell is approximately 100 times thinner than a human hair. The amyloid deposition, or plaque, is a neuropathological hallmark of the disease.

Below: Individual nerve cell processes (green) passing by toxic amyloid deposits (red) in a living mouse brain. The nerve cell appendages closest to the deposits are swollen, showing early signs of degeneration. Photos by Julia Tsai and Wen Biao Gan.

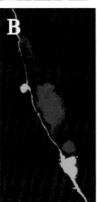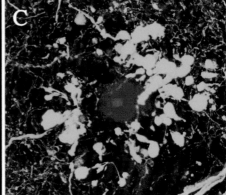

occurs with Golgi's technique. The fluorescent protein method also has the significant advantage of allowing scientists to observe living brain cells as they undergo growth and death. The researcher in the Prologue, for example, used fluorescent proteins to visualize directly living neurons and witness them decay and become consumed by plaques during the pathological development of Alzheimer's disease.[12] The ability to label specific cells allows researchers to view the specific activities of brain cells.

An area of neuroscience research that has been significantly advanced by the visualization power of fluorescent proteins is the study of smell. It is known that the olfactory system detects over 500,000 different scents, and provides the foundation of taste. The tongue produces only five basic sensations: salt, sweet, sour, bitter, and umami (a meaty or savory flavor). While a person is chewing, molecules of food become airborne and drift into the nose through a passageway in the back of the mouth. Deep in the back of the nose, yet still exposed to the air, resides a patchwork of millions of olfactory nerve cells that sense odor. Most smells are mixtures of hundreds to thousands of different volatile molecules at various concentrations. It remains a mystery how the olfactory system detects and discriminates thousands of different chemicals. Some believe a molecule's shape determines its smell, while others postulate that smell receptors in the nose detect intramolecular vibrations of odor molecules.

In 1991, two molecular biologists at Columbia University, Linda Buck and Richard Axel, discovered a collection of genes encoding membrane proteins.[13] The new group consisted of families of closely related genes, often found seated next to each other on a chromosome. The genes they discovered, approximately 1,000, remain the largest collection of closely related genes in humans. Buck and Axel began to suspect that they had

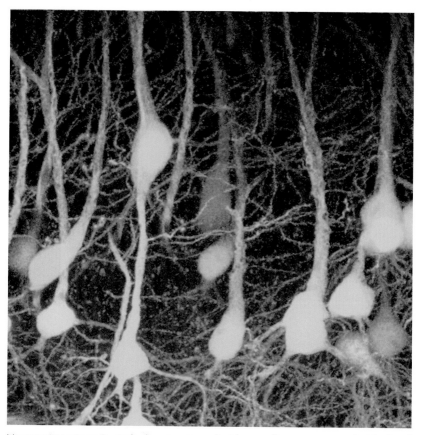

Neurons in a mouse's cerebral cortex tagged with cyan fluorescent protein (blue) and yellow fluorescent protein (green). Courtesy of Jeffery Lichtman.

found the elusive olfactory receptor proteins and sought to determine if they were present in olfactory neurons in the nose.

A former member of Axel's laboratory and now director of an olfactory group at the Rockefeller University, Peter Mombaerts, linked the green fluorescent protein gene to one of their newfound proteins. He then inserted this gene into a mouse. When he looked under the microscope into the animal's nose, only the nerve cells with that specific receptor glowed bright green. The axon from each cell could be easily tracked as it passed from the nose, through a tiny hole in the skull, and into the brain. The glowing cells were randomly spread throughout the nose, but the researchers were surprised to discover that all the glowing axons converged at a single spot in the olfactory bulb. These bulbs are slender pill-shaped structures in the front of the brain—in humans, each the size of a Tic-Tac. Spread over the surface of each bulb are round structures, giving them a leopard-skin appearance.

Every newborn nerve cell in the nose sends out an axon that slinks through the skull into the brain and connects to its exact position on the appropriate olfactory bulb. To put things into perspective, if an olfactory nerve cell was the size of a person and the person's arm was the axon, the hand would have to blindly navigate the length of a football field through a crowd of thousands of people and come to rest on the shoulder of a specific person. With every sneeze, many of these cells are killed, and new nerve cells in the nose quickly replace dead ones and their axons must embark on the same journey into the brain. How the neurons perform this pilgrimage is not understood, but it is essential for scent discrimination. To further complicate matters, olfactory neurons are virtually identical to one another. The only difference between neurons is which receptor protein the cell produces and exports to its surface. The

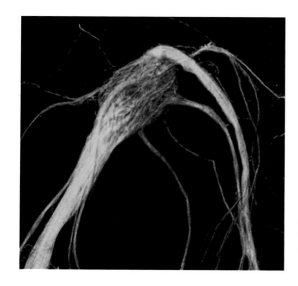

GFP-tagged mouse olfactory receptor axons. Courtesy of Peter Mombaerts.

receptor protein is where the odor molecule attaches and it also governs how the neuron connects with other neurons in the brain.

By lighting up individual olfactory receptor proteins with fluorescent proteins, the researchers could assure that only those olfactory neurons glowed green. They could track the radiant axon of a single olfactory receptor cell into the brain and witness where it connected. These experiments revealed a staggering and novel organization scheme that required scientists to rethink how the olfactory system functions. The scent of chamomile, for example, activates one set of olfactory receptor neurons, while saffron activates another set; some neurons of the two sets may overlap, but most of them do not. The brain somehow perceives the pattern of receptor neuron activation and decides which scent is being smelled. The Nobel Prize in Physiology or Medicine in 2004 was awarded to Axel and Buck (almost a century after Cajal and Golgi) for their discoveries of odorant receptors and the organization of the olfactory system. Fluorescent proteins were featured prominently, present in 23 slides of Axel's 45-minute Nobel lecture.

CHAPTER TWELVE

Glowing Thoughts

DURING THE first half of the twentieth century, scientists built on Cajal's neuron theory and determined that the brain consists of a staggering number of neurons in a cornucopia of shapes and sizes. It became evident that neurons huddle together in high densities in groups dedicated to specific tasks such as smell, vision, and hearing. The general function of these areas of the brain has been identified, but the inner workings remain largely a mystery.

Early work to map the brain's functional connections was begun in the 1930s by a Swiss physiologist, Walter Rudolph Hess. Using mainly cats, he stimulated precise areas of the brain with low levels of electrical current. Hess discovered that these tiny shocks, barely enough to cause a tingle if delivered to the finger, resulted in dramatic behavioral and physiological responses. "On stimulation within a circumscribed area of the ergotropic-dynamogenic zone, there regularly occurs . . . a manifest change in mood. Even a formerly good-natured cat turns bad-tempered; it starts to spit and, when approached, launches a well-aimed attack," wrote Hess. By stimulating other areas of the brain, he found different sorts of responses: "The blood pressure, for example, does not respond by a rise, but by a fall; the heart rate does not increase, but rather decreases. At the same time respiration slows down, as opposed to the speeding-up which is obtained from the ergotropic-dynamogenic zone."[1] Hess's work revealed that the elaborate neuronal structures discovered by Cajal used electrical signals to communicate with the body.

After World War II, scientists took advantage of the sensitive electronic components developed for radar and sonar to measure and manipulate these electrical signals found in nerve cells. They discovered that neurons generated their own electricity to communicate by concentrating potassium ions inside the cell, while excluding sodium ions. This is accomplished via selective membrane tunnels, known as ion channels, which penetrate the cell—some specific to potassium, others to sodium. The tunnels act as molecular gatekeepers, similar to bouncers at an exclusive nightclub deciding who gains entry and who is excluded. Since ions are charged molecules, their movement and unequal concentration on either side of the cell creates a small voltage. This voltage gives each nerve cell the property of a battery. When the cell's voltage reaches a critical threshold, the sodium channels snap open, causing a flood of sodium ions to rush into the cell. This flood of positively charged ions causes a brief spike in the cell's voltage. The spike, or action potential, spreads, like a ripple in a pond, across the surface of the neuron. When the voltage ripple reaches one of the nerve endings, it causes a squirt of potent chemicals, called neurotransmitters, onto neighboring neurons. Neurotransmitters conduct the electrical signal across the gap of the synapse to the next nerve cell. These scientific studies of the electrical properties of neurons gave life to the structures Cajal had identified. The brain is composed of massively interconnected nerve cells, communicating via a staccato of electrical and chemical pulses. Somewhere within them lies human consciousness. Understanding and interpreting this neuronal firing both is difficult and presents researchers with a range of ethical dilemmas.

José Manuel Rodriguez Delgado, a flamboyant Spanish-born neuroscientist active from the 1950s to the 1970s, did not always deal satisfactorily with these dilemmas. Delgado began stimulating different areas of the brain and monitoring the behavioral effects. Although many other scientists were doing similar research, his flair and predilection for high-profile stunts and experiments kept him in the public eye. He was an ardent proponent of direct electrical brain stimulation to explore the cerebral bases of anxiety, pleasure, and aggression, and also suggested that it could be used to treat human behavioral "defects." In popular presentations of his work, he larded discussions of valid therapeutic goals with ideas of using brain implants to manipulate "antisocial" human behaviors. He also publicly defended the large-scale use of such technologies by governments and societies to affect social development, equating such behavioral modifications to a society's use of schooling and laws to modify its citizens' behavior. Delgado, who was trained at Madrid School of Medicine, was inspired by Cajal's legacy. "Cajal said that knowledge of the physicochemical basis of memory, feelings, and reason would make man the true master of creation, that his most transcendental accomplishment would be the conquering of his brain," wrote Delgado.[2]

In 1964, Delgado, then a professor at the Yale University School of Medicine, demonstrated brain stimulation in a bullfighting arena in Cordova, Spain. He used a "Stimoceiver," a radio transmitter connected to a wire electrode implanted in a bull's brain. The next morning, upon recovery from the implantation, the bull was led into the ring and taunted by a matador with a red cape. Several moments later, Delgado slipped into the ring wearing gray slacks, a pullover sweater, and a black tie instead of the traditional matador outfit.[3]

José M. R. Delgado halting a charging bull with remote-controlled electrodes in the bull's brain, 1964. Courtesy of José M. R. Delgado.

Delgado held a bulky remote control in one hand. The angered bull noticed him and began to charge. Just moments before the bull could gore Delgado, he pressed the remote, activating the Stimoceiver and stimulating the bull's brain. The bull came to a halt. Finally Delgado pressed another button, stimulating a different area of the bull's brain, and the bull obediently turned and trotted away.

The *New York Times* heralded the event in a front-page article, headlined "Matador with a Radio Stops Wired Bull," as "probably the most spectacular demonstration ever performed of the deliberate modification of animal behavior through external control of the brain."[4] In Delgado's 1969 book, *Physical Control of the Mind: Toward a Psychocivilized Society*, he wrote: "It was also repeatedly demonstrated that cerebral stimulation produced inhibition of aggressive behavior, and a bull in full charge could be abruptly stopped. The result seemed to be a combination of motor effect, forcing the bull to stop and to turn to one side, plus behavioral inhibition of the aggressive drive. Upon repeated stimulation, these animals were rendered less dangerous than usual, and for a period of several minutes would tolerate the presence of investigators in the ring without launching any attack."[5]

Delgado did not confine his work to animals. He also described experiments performed on humans. These studies were always performed with the intention of helping patients with chronic neurological conditions. Delgado developed a procedure in which several stainless steel wires, 0.1 millimeter in diameter, are introduced through a hole through the skull and into the brain.[6] In a chapter of *Physical Control of the Mind* entitled "Hell and Heaven within the Brain," he described the range of emotional responses to electrical stimulation of specific areas of the brain. In one case a female patient reacted violently to brain stimulation: "A 1.2 milliampere excitation of this point [of the brain] was applied while she was playing the guitar and singing with enthusiasm and skill. At the seventh second of the stimulation, she threw away the guitar and in a fit of rage launched an attack against the wall and then paced around the floor for several minutes, after which she gradually quieted down and resumed her cheerful behavior. This effect was repeated on two different days."[7]

By implanting wires and passing low levels of electrical current, Delgado and other neurophysiologists "mapped" behavioral responses when different brain regions were stimulated. Activation of different areas induced feelings of pleasure, pain, numbness, anger, fear, and rage. Other areas of the brain, when stimulated, elicited experiences such as perceptions of flashes of colors and shapes, perception of sounds, and tactile sensations.

In the 1970s, this type of research began to provoke a strong backlash. On February 24, 1972, Cornelius Gallagher, a U.S. representative from New Jersey, stood before the House of Representatives and submitted what he called "one of the most shocking documents I have ever seen," 11 pages written by Peter R. Breggin, a psychiatrist, entitled "The Return of Lobotomy and Psychosurgery." The goal was to alert the American public to what Gallagher viewed as Orwellian brain experiments.[8] Breggin wrote:

> Delgado is working on the ultimate lobotomy—direct long term physical control of human beings . . . Despite his denials that there is anything reminiscent of *1984* about all this, he has been working on remote control of humans by computers which can selectively inhibit various emotions as they are detected and recorded from brain waves . . . While this is "speculative," it is by no means a remote possibility. If a few men can do what they have done working in isolated labs with little financial support, they might in a crash program develop complete computerized control of humans in a matter of years.[9]

Breggin went on to say that Delgado envisioned armies of electrode-implanted soldiers controlled by generals, and that Delgado thought violent or socially unacceptable people could be controlled by brain implants.

Delgado tried to address the fears of people like Breggin by posing a question: "Could a ruthless dictator stand at a master radio transmitter and stimulate the depth of the brains of a mass of hopelessly enslaved people?" He admitted that "it is true that we can influence emotional reactivity and perhaps make a patient more aggressive or amorous," but, he added, "by means of ESB [electrical stimulation of the brain], we cannot substitute one personality for another, nor can we make a behaving robot of a human being."[10]

Complete computerized control of human behavior is not yet a reality, or even a scientific focus. But there has been a recent research surge in direct electrical interfacing of the brain. There is a little-known appendage of the Pentagon called the Defense Advanced Research Projects Agency, or DARPA, that has been a strong supporter of this type of work. DARPA was established in 1958 by President Dwight D. Eisenhower as the U.S. response to the Soviet Union's surprise launch of the first space satellite, *Sputnik*. In its short history, DARPA's successes include creating the Internet, highly effective night-vision goggles, and radar-evading stealth aircraft. DARPA's mission is to "maintain the technological superiority of the U.S. military and prevent technological surprise from harming national security by sponsoring revolutionary, high-payoff research that bridges the gap between fundamental discoveries and their military use."[11]

In 2002, DARPA-funded scientists at the State University of New York in Brooklyn performed experiments in which a rat, having had stimulation electrodes implanted in its motor cortex, could be directed to navigate through mazes.[12] The researcher could send real-time commands from a joystick to "Roborat," instructing it to move forward, turn left or right, and stop. Roborat inspired DARPA, which began to envision a future in which the brains of soldiers could be directly tapped to control their actions, communicate with command structures, and directly interface with weapon systems. DARPA believes that soldiers of the twenty-first century could be outfitted with devices that directly communicate thoughts, bypassing the physical actions generally associated with them. "In the long run, we could have brain-to-brain communication; we could improve the performance of normal healthy individuals," says Alan Rudolph of DARPA.[13] DARPA is also engaged in research aimed at influencing brain function by directly tapping into the electrical circuitry of the brain. Such investigation of a two-way interaction with the brain, sponsored by an arm of the military, conjures up frightening images of mind control.

Like stem cell research and human cloning, this type of research is not always widely embraced by the public. There is the belief, as voiced by Breggin, that soon it will be possible to produce soldiers or agents whose behavior, witting or unwitting, is programmed or controlled externally. Could brain implantation be used to force subjects to commit inappropriate, dangerous, or illegal acts? Such ideas may seem to apply only in science fiction and popular movies such as *The Manchurian Candidate* or *The Matrix*, but as the brain becomes less of a mystery, anything is possible. It is essential to proceed with openness and ethical guidance. But the potential benefits of brain-machine interfacing are enormous. It offers the hope of finding treatments for some of the most intractable and devastating neurological conditions that plague us, such as epilepsy, paralysis, chronic pain, and even blindness.

At 10 o'clock on the night of July 3, 2001, Matthew Nagle was watching the fireworks display on Wessagussett Beach in Weymouth, Massachusetts, an annual tradition for the small New England town. Suddenly a fight erupted. Nagle, a 21-year-old 195-pound former linebacker—and record holder for unassisted tackles at Weymouth High School—clawed to the center of the melee, hoping to assist friends entangled in the brawl. The last thing Nagle recalls is someone screaming about a knife. Nagle fell heavily to the ground with an 8-inch curved knife lodged in his upper spine. When the ambulance arrived, Nagle's heart was not beating, but he was resuscitated by paramedics.[14]

Nagle survived, but his spinal cord, the nerve tissue that connects the brain to the body, was severed. Nagle checked into several rehabilitation centers in the 3 years following the stabbing, but his injury left him paralyzed from the neck down. He is unable to control his respiration and must rely on a ventilator for every breath. Although Nagle is fully conscious and can move his face, he has no feeling in the rest of his body. Unfortunately, there is currently no treatment for his condition. Nerve cells do not regenerate and, over time, scars form around the once-connected nerve fibers, further reducing any chance of improvement.

Following the attack, Nagle moved back to his parent's house in Weymouth. His father, a homicide detective for the Cambridge, Massachusetts, police force, took a leave of absence from his job and his mother dropped out of graduate school to help care for him. But despite their efforts, Nagle sank into a severe depression and often contemplated suicide. Doctors told him that he might never move again.

Currently there are no effective surgeries or drug treatments for such injuries. Rehabilitation provides only modest improvements and only in patients with partial damage. Victims of spinal cord damage, more than 250,000 people in the United States alone, are sentenced to life imprisonment in a body that cannot be controlled and has no feeling.[15]

The business card of Cyberkinetics, Inc., reads "Turning Thought into Action." It may sound like a typical corporate logo, but this biotechnology company created in 2001 by John Donoghue, a Brown University neuroscience professor, takes the saying literally. At first the company, based in Foxboro, Massachusetts, consisted mainly of members of Donoghue's laboratory, working to detect and interpret neural signals. In late 2002 Cyberkinetics merged with Bionic Technologies, a manufacturer of neural recording equipment, and soon the new company began developing the first brain-machine interface technology capable of implanting electrode arrays into the human brain.

The implant is a square array of 125 electrodes, resembling a miniature bed of nails the size of a baby aspirin. This device is surgically implanted under the skull in the motor cortex, a portion of the brain involved in body movements. Once embedded, it is capable of monitoring the electrical activity of hundreds of individual brain cells simultaneously. Only a small connector port is visible on surface of the scalp. While the electrodes inside the brain listen to the brain's chatter, complex computer algorithms instantly translate the patterns of neuronal firing and decipher their intent. Neurons in the motor cortex, for example, fire in a specific sequence to move the arm up and down. Once these patterns are decoded, the computer can command robotic appendages to enact the thought of arm movement—bypassing the brain message to the spinal cord and the body's muscles.

In 2004 Matthew Nagle became the first human recipient of the experimental implant device called BrainGate. He had heard about the new technology at a hospital and his mother contacted Cyberkinetics. "If I can't help myself, through this research, I can at least help someone else," says Nagle. The electrode array was inserted into Nagle's motor cortex during a 6-hour operation on June 22, 2004. Five months later, after Nagle's headaches had subsided and the wound had stopped draining, the experiments began. Nagle participated in many experiments where he found himself sitting in a darkened room staring at a computer screen. He was instructed to follow a red dot dancing about—making circular and zigzag patterns with his eyes. Behind him sat a Cyberkinetics scientist, surrounded by a bank of computer screens outside Nagle's view. On one screen, the scientist watched the same moving red dot, but he also saw a green dot loosely trailing the red dot. The green dot, which Nagle could not see, was directly controlled by his thoughts. When the red dot moved to the left, Nagle willed his body in that direction. The computer analyzed the activity of his brain cells and converted intended motions into movements of the green dot. Several weeks later, Nagle had progressed to playing the video game *Tetris* with his thoughts. Soon he hopes to advance to other games. Cyberkinetics is customizing the software to Nagle's wishes, but their overall goal is to develop brain-machine interfacing to a point where people who no longer have the ability to move by their own volition can control prosthetic devices.

Electrode-recording experiments, such as those being carried out on Nagle, are tantalizing and provide a glimpse of positive potential for this type of research. But they are a long way from providing victims with independence beyond simple movements. Implant devices have crippling limitations that will probably not allow direct monitoring of enough brain cells to provide adequate information to perform more complex actions. Even simple movements of the arms and legs require the concerted activity of millions of brain cells acting in unison. Consider the relatively effortless process of reaching for a glass of water. Before the action takes place, the brain produces an ordered list of muscle commands. This list is then delivered to the spinal cord, where motor neurons activate the appropriate muscles in sequence. During the process, feedback from receptors in the muscles notifies the brain and spinal cord of the status of the process and, if necessary, the motion is corrected.

In the case of Matthew Nagle, the brain generates the commands for muscle movement, but the commands never reach the spinal cord and motor neurons. To reroute the signal around Nagle's injury, scientists must record the activity of the specific brain cells, of the billions present, that produce motor commands. The electrical chatter of the neurons, which sounds like a rapidly firing machine gun, then needs to be translated into meaningful motor commands. At this stage, a computer can be programmed to direct a robotic arm to carry out the intended motions.

To describe movements, it is necessary to record the activity of individual nerve cells rather than large groups. In order to monitor directly the electrical activity of a single neuron, thin insulated wires, half the width of a human hair, are positioned in the brain. These electrodes detect the electrical activity of nerve cells that contact the exposed tip. These skinny wires are still hundreds of times larger than each nerve cell, which makes the process akin to dropping a microphone the size of a wrecking ball into a crowded skyscraper in order to hear what one person on the fifth floor is saying. During the implantation of the electrodes, many more nerve cells are damaged than are recorded. When neurons are damaged, the surrounding tissue reacts, engulfing the electrode in scar tissue, and eventually silencing it. Recording from a large number of neurons presents a logistical dilemma. Nerve cells in the human cerebral cortex are packed together at a dense congregation, tens of thousands residing in an area the size of a pinhead. Wires cannot be created that are fine enough to record from each neuron, and increasing the number of wires in a particular region damages more of the overlying cells. At a certain threshold, each additional wire leads to diminishing returns. The future of brain-machine interfacing is impeded by the inability to monitor enough brain cells to interpret intended actions. One of the founders of Cyberkinetics, Mikhail Shapiro, believes that "it is going to be really hard to get a combination of large-scale and high-resolution imaging with electrodes," and says that in the future, "you are going to have to move to something like optics."[16]

Enter fluorescent proteins.

In 1997, Ehud Isacoff, a professor of neurobiology at the University of California, at Berkeley, and a graduate student, Micah Siegel, wanted to understand how potassium ion channels operate. When neurons are electrically active, subtle changes in conformation open the channel and allow potassium to exit the cell. To examine the

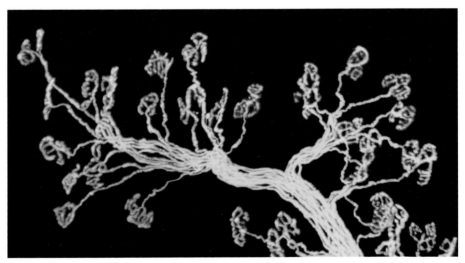

Nerve fibers from neurons that contain the gene for yellow fluorescent protein. The fibers travel together in a bundle from the spinal cord and then branch to end on individual muscle fibers. The circular red endings are the synapses between the nerve fibers and the muscles. Electrical signals travel down the axons and cause the muscle to contract. Photo by Jeffery Lichtman.

change in the shape of the potassium channels, they decided to illuminate them by inserting the sequence for green fluorescent protein. When they plugged this new glowing potassium channel into cells it had an unexpected property: the cell changed its fluorescence intensity when the cell's membrane voltage changed.[17] This phenomenon seemed odd. The reliable stability of fluorescent proteins is a primary reason they have become popular tools. The reason the fluorescence changes is still not fully explained, but Isacoff and Siegel thought they had made an important discovery. Inadvertently, they had stumbled on a probe that converts changes in a cell's membrane voltage into an optical signal.

Siegel and Isacoff's probe excited the neuroscience community, but ultimately it became evident that it could not be used in nerve cells. First, there is a delay in the time from when the cell changes voltage to when the fluorescence changes. Also, although it works well in frog eggs, used because of their large size, it doesn't work in brain cells, which are much smaller.

One of the authors of this book (Pieribone) was inspired by Siegel and Isacoff's research and began developing a new probe at the Yale University School of Medicine. Like most neurobiologists, Pieribone used electrodes to study cell activity. He was frustrated that the methods used to study the electrical activity in brain cells had not improved since their inception about 50 years earlier. Neurophysiology is one of the few areas of science that had not benefited from the molecular biology revolution.

Trained as both a physiologist and a molecular biologist, Pieribone sought a better way to study how brain chatter produces thoughts and actions. After 4 years of research, Pieribone and his graduate student, Kazuto Ataka, unveiled a second fluorescent probe in 2002.[18] They created a voltage-sensitive fluorescent ion channel that responded a hundred times more rapidly to changes in membrane voltage than that of Siegel and Isacoff. But like the first probe, it does not function in nerve cells. Pieribone and Isacoff have teamed up with a diverse group of scientists in an effort to develop a more useful probe.

It is only a matter of time before a fluorescent probe is created that responds to changes in electrical activity and also works inside the living brain. The payoff will be tremendous: a technique that will provide scientists with a way to explore the brain unobtrusively. Researchers could target these glowing probes to specific subsets of nerve cells, making it possible to record a single nerve cell firing. In this fashion, the activity of hundreds, thousands, and even millions of neurons could be recorded simultaneously without injuring the brain tissue. This optical signature would improve brain-machine interfacing.

In the future, a brain-machine interface may consist of a tiny camera placed under the scalp of a paralyzed person. This camera would image fluorescent neurons on the surface of the brain and record changes in their fluorescent intensity. The individual neuronal spikes would be collected from the different cells in the region as changes in a fluorescent signal. This information would then be fed into a sophisticated computer. The coincident activity and firing relationships between the cells would be compared during motor movements. The researcher could ask the subject to reach for a glass and then record how the neurons fired to create this action. The larger the range of motions to be discriminated, the more cells it is necessary to record. An algorithm is then developed that can interpret the complex patterns of neuronal activity that signal various motions. Generating such an algorithm would require training periods during which the computer would learn how to interpret the seemingly random neuronal activity to create motor commands. These

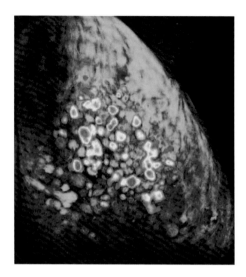

The response of odorant receptors on the olfactory bulb of a mouse to the odor of methyl benzoate, which smells like wintergreen. The colored spots are areas where there was a change in the fluorescent protein's intensity, indicating neuronal activity.

commands could then be directed either to robotic limbs to carry out the actions or, some day, to electrodes in the musculature that could electrically stimulate muscles in defined patterns to produce body motions.

In this fashion, people with spinal cord damage that disconnects the brain from its targets in the body's musculature could be reconnected through a computer interface. Such interfacing requires high-speed computers that can process hundreds to thousands of fluorescent images per second of the brain's surface and then rapidly respond with the correct output in real time. Instead of placing electrodes next to individual nerve cells to listen to what they are saying, this interface would convert what they are saying into a visible signal capable of being monitored from afar. The more shifted the fluorescent protein is toward the red side of the spectrum, the farther the signal will penetrate through the brain. Once this procedure is refined, surgically implanted electrodes will become obsolete and the number of nerve cells that can be monitored will jump from 125, as in Nagle's case, to millions. With the success of electrical brain-machine studies such as Nagle's, one can imagine what a ten thousand–fold advance in resolution could bring.

How might such an image of the brain look? Recent studies in the olfactory bulb of the mouse provide a glimpse. Gero Miesenböck and James Rothman at Memorial Sloan Kettering Cancer Institute created a hybrid fluorescent protein that also changes its fluorescence.[19] But instead of reporting changes in voltage, it alters its brightness when synaptic vesicles fuse and release neurotransmitters. Researchers have since inserted this hybrid fluorescent protein into individual mouse olfactory neurons. It is now possible to watch, in real time, different neurons in the nose become activated by distinct scents. The activity of a dynamic living brain can now be converted into light and made visible for the world to see.

Cajal once wrote that "countless modifications during evolution have provided living matter with an instrument of unparalleled complexity and remarkable functions: the nervous system, the most highly organized structure in the animal kingdom."[20] It is in this nervous system that we now use fluorescent proteins to look inward to understand the greatest mystery of human existence, our consciousness. Fluorescent proteins, originally plucked from marine creatures, are a product of diverse life forms that have existed and evolved for millions of years. Scientists have only begun to tap this biological diversity for tools such as fluorescent proteins that can be used to cure ailments, investigate pressing biological questions, and resolve the great enigmas of human life. As Osamu Shimomura said in August 2004 at a conference in his honor when he was asked why years ago he was scooping jellyfish out of the water: it was "to solve a mystery."[21]

NOTES • ACKNOWLEDGMENTS • INDEX

Notes

1. Living Light

1. Edith A. Widder, "Bioluminescence and the Pelagic Visual Environment," *Marine and Freshwater Behaviour and Physiology* 35 (2002): 1–26; Edmund Newton Harvey, *Living Light* (New York: Hafner Publishing Co., 1965).
2. Aristotle, *On Colours*, in *Aristotle: Minor Works*, trans. W. S. Hett (Cambridge, Mass.: Harvard University Press, 1936), p. 7.
3. Aristotle, *De Sensu and De Memoria*, trans. G. R. T. Ross (Cambridge: Cambridge University Press, 1906), pp. 47, 49.
4. Pliny, *Natural History*, vol. 3, books 8–11, trans. H. Rackham, Loeb Classical Library (Cambridge: Harvard University Press, 1940), p. 287.
5. George Sarton, *Introduction to the History of Science*, vol. 3: *Science and Learning in the Fourteenth Century*, (Baltimore: Williams & Wilkins, 1947), p. 488.
6. *The Divine Comedy of Dante Alighieri*, trans. Henry F. Cary, vol. 20 (New York: P. F. Collier, 1909), p. 109.
7. J. Macartney, "Observations upon Luminous Animals," *Philosophical Transactions of the Royal Society of London* 100 (1810): 274.
8. Raphaël Dubois, "Note sur la fonction photogénique chez les Pholades," *Comptes Rendus des Séances de la de Société Biologie* (Paris) 39 (1887): 566.
9. Raphaël Dubois, "Les elaterides lumineux," *Bulletin de la Société Zoolologique de France* 11 (1886): 1–275.
10. David Cecil Smith and Angela Elizabeth Douglas, *The Biology of Symbiosis* (London: Edward Arnold, 1987), p. 224.

11. John Thomas Osmond Kirk, *Light and Photosynthesis in Aquatic Ecosystems* (Cambridge: Cambridge University Press, 1983).

12. Ole Munk, "Histology of the Fusion Area between the Parasitic Male and the Female of the Deep-sea Anglerfish, *Neoceratias spinifer* Pappenheim, 1914 (Teleostei, Ceratiodei)," *Acta Zoologica* (Stockholm) 81 (2000): 315–324.

13. Stephen Jay Gould, *Hen's Teeth and Horse's Toes* (New York: W. W. Norton, 1983), p. 31.

14. Charles Darwin, *The Voyage of the H.M.S. Beagle* (New York: D. Appleton and Co., 1890), p. 173.

15. Paul König, *Voyage of the Deutschland: The First Merchant Submarine* (New York: Hearst's International Library Co., 1917), p. 111.

16. Nikolai Ivanovich Tarasov, "Marine Luminescence," trans. U.S. Naval Oceanographic Office from *Svyecheniye Morya* (Moscow: USSR Academy of Sciences, 1956), p. 15.

17. Ibid., p. 20.

18. James A. Lovell and Jeffrey Kluger, *Lost Moon: The Perilous Voyage of Apollo 13* (New York: Houghton Mifflin, 1974), pp. 68–69.

19. Interview with Mark Moline, Feb. 20, 2005.

20. Ocean Studies Board, Commission on Geosciences, Environment, and Resources, "Oceanography and Naval Special Warfare: Opportunities and Challenges," (Washington, D.C.: National Academy Press, 1997), p. 31.

2. Fireflies of the Sea

1. Frank H. Johnson, "Edmund Newton Harvey, 1887–1959," *Biographical Memoirs, National Academy of Sciences USA* 39 (1967): 193.

2. Ibid., p. 195.

3. Ibid., p. 204.

4. Ibid., p. 216.

5. Ibid., p. 217.

6. Edmund Newton Harvey, "Studies on the Permeability of Cells," *Journal of Experimental Zoology* 10 (1911): 507–556.

7. Johnson, "Edmund Newton Harvey," p. 218.

8. Ibid., p. 195.

9. James Frederic Danielli and Edmund Newton Harvey, "The Tension at the Surface of Mackerel Egg Oil, with Remarks on the Nature of the Cell Surface," *Journal of Cellular and Comparative Physiology* 5 (1935): 483–494; James Frederic Danielli and Hugh Davson, "A Contribution to the Theory of Permeability of Thin Films," *Journal of Cellular and Comparative Physiology* 5 (1935): 495–508.

10. Edmund Newton Harvey, "On the Chemical Nature of the Luminous Material of the Firefly," *Science* 40 (1913): 33–34.

11. Ethel Browne Harvey, *The American Arbacia and Other Sea Urchins* (Princeton: Princeton University Press, 1965).

12. Johnson, "Edmund Newton Harvey," p. 220.

13. James G. Morin, "'Firefleas' of the Sea: Luminescent Signaling in Marine Ostracode Crustaceans." *The Florida Entomologist* 69 (1986): 105–121.

14. Edmund Newton Harvey, *Living Light* (New York: Hafner Publishing Co., 1965), pp. 301–302.

15. Frank H. Johnson, *Luminescence, Narcosis, and Life in the Deep Sea* (New York: Vantage Press, 1988), p. 7.

16. Harvey, *Bioluminescence*, p. 301.

17. Osamu Shimomura, personal communication.

18. H. Arthur Klein, *Bioluminescence* (Philadelphia: J. B. Lippincott Co., 1965), p. 119.

19. Frank H. Johnson, "Foreword," in Peter J. Herring, ed., *Bioluminescence in Action* (New York: Academic Press, 1978), p. viii.

20. Edmund Newton Harvey, "The Mechanism of Light Production in Animals," *Science* 44 (1916): 208–209.

21. Harvey, *Bioluminescence*, p. x.

22. Rupert S. Anderson, "The Partial Purification of *Cypridina* Luciferin," *Journal of General Physiology* 19 (1935): 301–305.

23. Harvey, *Bioluminescence*, p. 306.

24. Edmund Newton Harvey, *A History of Luminescence: From the Earliest Times until 1900* (Philadelphia: The American Philosophical Society, 1957).

25. Johnson, *Luminescence*, p. 48.

26. Ibid., p. 48.

3. From the Fires of Nagasaki

1. Unless otherwise indicated, information about Shimomura's time in Japan and quotations come from two interviews (Aug. 21, 2003, and Dec. 28, 2003) at his home, his unpublished recollections, and an article published in the *Cape Cod Times:* Osamu Shimomura, "Woods Hole scientist recalls day the bomb fell on Nagasaki," *Cape Cod Times,* Aug. 6, 1995, p. G1.
2. Robert E. Haney, *Caged Dragons: An American P.O.W. in W.W. II Japan* (Ann Arbor: Momentum Books, 1991), p. 131.
3. United States Strategic Bombing Survey, "Summary Report (Pacific War)" (Washington, D.C., July 1, 1946), p. 16.
4. Jim B. Smith and Malcolm McConnell, *The Last Mission: The Secret Story of World War II's Final Battle* (New York: Broadway Books, 2002), p. 99.
5. Richard B. Frank, *Downfall: The End of the Imperial Japanese Empire* (New York: Random House, 1989), p. 285.
6. United States Strategic Bombing Survey, "Summary Report," p. 24.
7. United States Strategic Bombing Survey, "Effects of the Atomic Bomb on Hiroshima and Nagasaki," ed. William Gannon (Santa Fe, 1973), p. 11.
8. Osamu Shimomura, "Discovery of Aequorin and GFP" *Journal of Microscopy* 217 (2005): 3–15.
9. Osamu Shimomura, Toshio Goto, and Yoshimasa Hirata, "Crystalline *Cypridina* Luciferin," *Bulletin of the Chemical Society of Japan* 30 (1957): 929–933.
10. Ibid.
11. Frank H. Johnson, "Light without Heat," *Princeton Alumni Weekly,* Nov. 29, 1976.

4. The Secret of the Jellyfish's Flicker

1. Osamu Shimomura, "Discovery of Aequorin and GFP," retirement talk at Woods Hole Marine Biological Laboratory, Woods Hole, Mass., June 27, 2002.
2. There is disagreement over the exact name of the species, and it is sometimes called *Aequorea aequorea, Aequorea forskalea,* or *Aequorea victoria.* Osamu Shimomura, personal communication.

3. Shimomura, "Discovery of Aequorin and GFP."

4. Ibid.

5. See the *Time* magazine cover article on Dixy Lee Ray, Dec. 12, 1977.

6. Edmund Newton Harvey, "Studies on Bioluminescence. XIII. Luminescence in the Coelenterates," *Biological Bulletin*, 41 (1921): 280–287.

7. Edmund Newton Harvey, *Bioluminescence* (New York: Academic Press, 1952).

8. Osamu Shimomura, "A Short Story of Aequorin," *Biological Bulletin* 189 (1995): 2.

9. Osamu Shimomura, "Discovery of Aequorin and GFP," lecture at the conference Calcium-regulated Photoproteins and Green Fluorescent Protein, Friday Harbor Laboratories, Wash., Aug. 29, 2004.

10. Osamu Shimomura, "Discovery of Aequorin and Green Fluorescent Protein," *Journal of Microscopy* 217 (2005): 9.

11. Shimomura, "Discovery of Aequorin and GFP," Friday Harbor lecture.

12. Osamu Shimomura, Frank H. Johnson, and Yo Saiga, "Microdetermination of Calcium by Aequorin Luminescence," *Science* 140 (1963): 1339–1440.

13. Ellis B. Ridgway and Christopher C. Ashley, "Calcium Transients in Single Muscle Fibres," *Biochemical and Biophysical Research Communications* 29: (1967): 229–234.

14. Osamu Shimomura, Frank H. Johnson, and Yo Saiga, "Extraction, Purification, and Properties of Aequorin, a Bioluminescent Protein from the Luminous Hydromedusan, *Aequorea*," *Journal of Cellular and Comparative Physiology* 59 (1962): 228.

15. D. Davenport and J. A. C. Nicol, "Luminescence in Hydromedusae," *Proceedings of the Royal Society of London* B 144 (1955): 399–411.

16. Shimomura, "Discovery of Aequorin and GFP," Woods Hole talk.

5. The Light at the End of the Rainbow

1. Lawrence Humphry, "Notes and Recollections," in Joseph Lamor, ed., *Memoir and Scientific Correspondence of the Late Sir George Gabriel Stokes* (Cambridge: Cambridge University Press, 1907), vol. 1, p. 7.

2. George Gabriel Stokes, "On Some Cases of Fluid Motion," *Transactions of the Cambridge Philosophical Society* 8 (1843): 105–137.

3. Lord Kelvin, "The Scientific Work of Sir George Stokes" (obituary notice), *Nature* 67 (1903): 337.

4. George Gabriel Stokes, "On the Theories of the Internal Friction of Fluids in Motion and of the Equilibrium and Motion of Elastic Solids," *Transactions of the Cambridge Philosophical Society* 8 (1845): 287–347.

5. George Gabriel Stokes, "On the Effect of the Internal Friction of Fluids on the Motion of Pendulums," *Transactions of the Cambridge Philosophical Society* 9 (1850): 8.

6. George Gabriel Stokes, "On the Change of the Refrangibility of Light," *Philosophical Transactions and Mathematical and Physical Papers* 3 (1852): 259.

7. George Gabriel Stokes, "Dynamical Theory of Diffraction," *Transactions of the Cambridge Philosophical Society* 2 (1849): 243–328.

8. Lord Kelvin, "The Scientific Work of Sir George Stokes," p. 337.

9. George Gabriel Stokes, "A Discovery," in Lamor, ed., *Memoir and Scientific Correspondence of the Late Sir George Gabriel Stokes*, vol. 1, p. 9.

10. Stokes, "On the Change of the Refrangibility of Light."

11. Sir Isaac Newton, *Opticks; or, A Treatise of the Reflections, Refractions, Inflections & Colours of Light* (London: Printers to the Royal Society, 1704).

12. Hugh O. McDevitt, "Albert Hewett Coons, June 28, 1912–September 30, 1978," *Biographical Memoirs, National Academy of Sciences* 69 (1996): 27–36.

6. Illuminating the Cell

1. James D. Watson and Francis H. C. Crick, "Molecular Structure of Nucleic Acids: A Structure for Deoxyribose Nucleic Acid" *Nature* 171 (1953): 737–738.

2. Osamu Shimomura, "Discovery of Aequorin and Green Fluorescent Protein," *Journal of Microscopy* 217 (2005): 11.

3. David Baltimore, "RNA-dependent DNA Polymerase in Virions of RNA Tumour Viruses," *Nature* 226 (1970): 1209–1211; Howard Martin Temin and Satoshi Mizutani, "RNA-dependent DNA Polymerase in Virions of Rous Sarcoma Virus," *Nature* 226 (1970): 1211–1213.

4. Interview with Douglas Prasher, Jan. 17, 2004, at the National Plant Germplasm Quarantine Center, Animal Plant Health Inspection Service, Beltsville,

Md. Unless otherwise indicated, this and further quotes from Prasher are from this interview.

5. Douglas Prasher, Richard O. McCann, and Milton J. Cormier, "Cloning and Expression of the cDNA Coding for Aequorin, a Bioluminescent Calcium-binding Protein," *Biochemical and Biophysical Research Communications* 126 (1985): 1259–1268.

6. John Blinks, GFP meeting, Friday Harbor, Washington, Aug. 2004.

7. James G. Morin and J. Woodland Hastings, "Energy Transfer in a Bioluminescent System," *Journal of Cell Physiology* 77 (1971): 313–318; J. Woodland Hastings and James G. Morin, "Calcium-triggered Light Emission in *Renilla*: A Unitary Biochemical Scheme for Coelenterate Bioluminescence," *Biochemical and Biophysical Research Communications* 37 (1969): 493–498.

8. H. Morise, Osamu Shimomura, Frank H. Johnson, and J. Winant, "Intermolecular Energy Transfer in Bioluminescent System of *Aequorea*," *Biochemistry* 13 (1974): 2656–2662; Osamu Shimomura, "Structure of the Chromophore of *Aequorea* Green Fluorescent Protein," *FEBS Letters* 104 (1979): 220–222.

9. William W. Ward, "Purification and Characterization of the Calcium-activated Photoproteins from the Bioluminescent Ctenophores, *Mnemiopsis* sp. and *Beroe ovata*" (Ph.D. diss., The Johns Hopkins University, 1974).

10. William W. Ward and Milton J. Cormier, "An Energy Transfer Protein in Coelenterate Bioluminescence: Characterization of the *Renilla* Green Fluorescent protein (GFP)," *Journal of Biological Chemistry* 254 (1979): 781–788.

11. Osamu Shimomura, "Discovery of Aequorin and GFP," retirement talk at Woods Hole, Mass., June 27, 2002.

12. William W. Ward, C. W. Cody, Russell C. Hart, and Milton J. Cormier, "Spectrophotomeric Identity of the Energy Transfer Chromophores in *Renilla* and *Aequorea* Green Fluorescent Proteins," *Photochemistry and Photobiology* 31 (1980): 611–615.

7. Glow Worms

1. Bob Edgar, "A Scientific Kokopelli," *Science* 294 (2001): 2103. This article also gives a description of the party after the conclusion of the Cold Spring Harbor Laboratory bacterial genetics course.

2. Sydney Brenner, as told to Lewis Wolpert, *Sydney Brenner: A Life in Science* (London: BioMed Central, 2001), p. 5.

3. Herbert George Wells, Julian S. Huxley, and George Philip Wells, *The Science of Life* (London: Cassell, 1931).

4. Brenner, *A Life in Science*, p. 6.

5. Sydney Brenner, "Autobiography," in *Les Prix Nobel, The Nobel Prizes, 2002*, ed. Tore Frängsmyr (Stockholm: Nobel Foundation, 2003), p. 270.

6. Ibid., p. 271.

7. Brenner, *A Life in Science*, p. 63.

8. Sydney Brenner, "Excerpts from Proposal to the Medical Research Council, October, 1963," In *The Nematode "Caenorhabditis elegans,"* ed. William B. Wood and the community of *C. elegans* researchers (Cold Spring Harbor: Cold Spring Harbor Laboratory Press, 1988), p. xii.

9. Sydney Brenner, Nobel lecture, "Nature's Gift to Science," in *Les Prix Nobel, The Nobel Prizes, 2002*, ed. Tore Frängsmyr (Stockholm: Nobel Foundation, 2003), p. 281.

10. Brenner, *A Life in Science*, p. 134.

11. Andrew Brown, *In the Beginning Was the Worm: Finding the Secrets of Life in a Tiny Hermaphrodite* (New York: Columbia University Press, 2003), p. 97.

12. John Sulston, "Autobiography," in *Les Prix Nobel, The Nobel Prizes, 2002*, ed. Tore Frängsmyr (Stockholm: Nobel Foundation, 2003), p. 357.

13. H. Robert Horvitz, "Autobiography," in *Les Prix Nobel, The Nobel Prizes, 2002*, ed. Tore Frängsmyr (Stockholm: Nobel Foundation, 2003), p. 307.

14. *The Worm Breeder's Gazette* was published from December 1975 through May 2003 (last issue). All issues are available online at: elegans.imbb.forth.gr/wli.

15. Bob Edgar, "The Shipping and Handling of Nematodes," *The Worm Breeder's Gazette* 1 (1975): 7.

16. John E. Sulston and H. Robert Horvitz, "Post-embryonic Cell Lineages of the Nematode, *Caenorhabditis elegans,*" *Developmental Biology* 56 (1977): 110–156.

17. Interview with Martin Chalfie, April 4, 2004. Unless otherwise indicated, further quotes from Chalfie are from this interview.

18. Interview with Douglas Prasher, Jan. 17, 2004, at the National Plant Germplasm Quarantine Center, Animal Plant Health Inspection Service,

Beltsville, Md. Unless otherwise indicated, further quotes are from this interview.

19. Douglas C. Prasher, Virginia K. Eckernrode, William W. Ward, Frank G. Prendergast, and Milton J. Cormier, "Primary Structure of the *Aequorea victoria* Green Fluorescent Protein," *Gene* 111 (1992): 229–233.

20. Marty Chalfie, Yuan Tu, and Douglas Prasher, "Glow Worms: A New Method of Looking at *C. elegans* Gene Expression," *The Worm Breeder's Gazette*, 13 (1993): 19.

21. Martin Chalfie, Yuan Tu, Ghia Euskirchen, William W. Ward, and Douglas C. Prasher, "Green Fluorescent Protein as a Marker for Gene Expression," *Science*, 263 (1994): 802–805.

8. Fluorescent Spies

1. Roger Tsien, "Unlocking Cell Secrets with Light Beams and Molecular Spies," Heineken Lecture, Sept. 23, 2002, Amsterdam, The Netherlands.

2. Ibid.

3. Oded Tour, René M. Meijer, David A. Zacharias, Stephen R. Adams, and Roger Y. Tsien, "Genetically Targeted Chromophore-assisted Light Inactivation," *Nature Biotechnology* 21 (2003): 1505–1508.

4. Interview with Roger Tsien, Jan. 5, 2004. Unless otherwise indicated, all further quotes from Tsien and information about him are from this interview.

5. Grzegorz Grynkiewicz, Martin Poenie, and Roger Y. Tsien, "A New Generation of Ca^{2+} Indicators with Greatly Improved Fluorescence Properties," *Journal of Biological Chemistry* 260 (1985): 3440–3450.

6. Douglas C. Prasher, V. K. Eckenrode, William W. Ward, Frank G. Prendergast, and Milton J. Cormier, "Primary Structure of the *Aequorea victoria* Green-fluorescent Protein," *Gene* 111 (1992): 229–233.

7. Satoshi Inouye and Frederick I. Tsuji, "*Aequorea* Green Fluorescent Protein: Expression of the Gene and Fluorescence Characteristics of the Recombinant Protein," *FEBS Letters* 341 (1994): 277–280.

8. Roger Heim, Douglas C. Prasher, and Roger Y. Tsien, "Wavelength Mutations and Post-translational Autooxidation of Green Fluorescent Protein," *Proceedings of the National Academy of Sciences* 91 (1994): 12501–12504.

9. Roger Heim, Andrew B. Cubitt, and Roger Y. Tsien, "Improved Green Fluorescence," *Nature* 373 (1995): 663–664.

10. U.S. patents 5,625,048, 5,777,079, 6,066,476, and 6,319,669 issued April 29, 1997, July 7, 1998, May 23, 2000, and Nov. 20, 2001 (respectively) to Tsien and Heim for Modified Green Fluorescent Proteins.

11. Hiroshi Morise, Osamu Shimomura, Frank H. Johnson, and J. Winant, "Intermolecular Energy Transfer in the Bioluminescent System of *Aequorea*," *Biochemistry* 13 (1974): 2656–2662.

12. Mats Ormö, Andrew B. Cubitt, Karen Kallio, Larry A. Gross, Roger Y. Tsien, and S. James Remington, "Crystal Structure of the *Aequorea victoria* Green Fluorescent Protein," *Science* 273 (1996): 1392–1395.

13. Ibid.

14. Interview with Andrew Cubitt, Jan. 4, 2004.

15. Interview with Pauk Berhm, May 7, 2004.

16. Trisha Gura, "Jellyfish Proteins Light Up Cells," *Science* 276 (1971): 189.

17. James G. Morin and J. Woodland Hastings, "Energy Transfer in a Bioluminescent System," *Journal of Cell Physiology* 77 (1971): 313–318.

18. Atsushi Miyawaki, Juan Llopis, Roger Heim, J. Michael McCaffery, Joseph A. Adams, Mitsuhiko Ikura, and Roger Y. Tsien, "Fluorescent Indicators for Ca^{2+} Based on Green Fluorescent Proteins and Calmodulin," *Nature* 388 (1997): 882–887.

19. Franck Mazyoer, "Le sacre des mutants," *Le Monde Diplomatique*, Jan. 2004, p. 28.

9. A Rosy Dawn

1. Interview with Sergey Lukyanov, March 22, 2004. Unless otherwise indicated, all quotes from Lukyanov and information about him are from this interview.

2. Sergey Lukyanov, "Cloning of a Novel Xenopus Homeobox Gene XANF-1 by Subtractive Hydridization" (Ph.D. diss., Moscow State University, 1985).

3. Interview with Mikhail Matz, March 22, 2004. Unless otherwise indicated, all quotes from Matz and information about him are from this interview.

4. Interview with Yulii Labas, March 22, 2004. Unless otherwise indicated, all quotes from Labas and information about him are from this interview.

5. Pushkin Fine Arts Museum director Irina Antonova in the Moscow News: english.mn.ru/english/issue.php?2003-10-19.

6. James G. Morin and J. Woodland Hastings, "Energy Transfer in a Bioluminescent System," *Journal of Cellular Physiology* 77 (1971): 313–318.

7. Interview with Andrey Romanko, March 22, 2004. Unless otherwise indicated, all quotes from Romanko and information about him are from this interview.

8. Interview with Sergey Fradkov, March 22, 2004. Unless otherwise indicated, all quotes from Fradkov and information about him are from this interview.

9. Mikhail V. Matz, Arkady F. Fradkov, Yulii A. Labas, Aleksandr P. Savitsky, Andrey G. Zaraisky, Mikhail L. Markelov, and Sergey A. Lukyanov, "Fluorescent Proteins from Nonbioluminescent Anthozoa Species," *Nature Biotechnology* 17 (1999): 969–973; Roger Y. Tsien, "Rosy Dawn for Fluorescent Proteins," *Nature Biotechnology* 17 (1999): 956–957.

10. Interview with Roger Tsien, Jan. 5, 2004.

11. C. E. S Phillips, "Fluorescence of Sea Anemones," *Nature* 119 (1927): 747.

12. Siro Kawaguti, "On the Physiology of Reef Corals. VI. Study on the Pigments," *Palao Tropical Biological Station Contributions* 2 (1944): 617–674.

13. René Catala-Stucki, *Carnival under the Sea* (Paris: R. Sicard, 1964), p. 4.

14. Charles H. Mazel, "Spectral Measurements of Fluorescence Emission in Caribbean Cnidarians," *Marine Ecological Progress Series* 120 (1995): 185.

15. Sophie G. Dove, M. Takabayashi, and Ove Hoegh-Guldberg, "Isolation and Partial Characterization of the Pink and Blue Pigments of Pocilloporid and Acroporid Corals," *Biological Bulletin* 189 (1995): 288–297.

16. Dmitry A. Shagin, Ekaterina V. Barsova, Yurii G. Yanushevich, Arkady F. Fradkov, Konstantin A. Lukyanov, Yulii A. Labas, Tatiana N. Semenova, Juan A. Ugalde, Ann Meyers, Jose M. Nunez, Edith A. Widder, Sergey A. Lukyanov, and Mikhail V. Matz, "GFP-like Proteins as Ubiquitous Metazoan Superfamily: Evolution of Functional Features and Structural Complexity," *Molecular Biology and Evolution* 21 (2004) 841–850.

10. Shimmering Reefs

1. Internet Movie Database Inc., www.imdb.com/boxoffice/alltimegross. Last update: April 11, 2005.

2. Clive Wilkinson, ed., *Status of Coral Reefs of the World: 2004* (Townsville: Australian Institute of Marine Science, 2004), p. 37.

3. "Flushing Nemo," *Harper's Magazine,* Oct. 1, 2003, p. 20.

4. Jörg Wiedenmann, Andreas Schenk, Carlheinz Röcker, Andreas Girod, Klaus-Diete Spindler, and G. Ulrich Nienhaus, "A Far-red Fluorescent Protein with Fast Maturation and Reduced Oligomerization Tendency from *Entacmaea quadricolor* (Anthozoa, Actinaria)," *Proceedings of the National Academy of Sciences USA* 99 (2002): 11646–11651.

5. Robert E. Campbell, Oded Tour, Amy E. Palmer, Paul A. Steinbach, Geoffrey S. Baird, David A. Zacharias, and Roger Y. Tsien, "A Monomeric Red Fluorescent Protein," *Proceedings of the National Academy of Sciences USA* 99 (2002): 7877–7882.

6. Marjorie L. Reaka-Kudla, "The Global Biodiversity of Coral Reefs: A Comparison with Rain Forests," in *Biodiversity II: Understanding and Protecting Our Biological Resources,* ed. Marjorie L. Reaka-Kudla, Don E. Wilson, and Edward O. Wilson (Washington, D.C.: Joseph Henry Press, 1997), pp. 83–108.

7. René L. A. Catala-Stucki, *Carnival under the Sea* (Paris: R. Sicard, 1964), p. 18.

8. Alfred H. Woodcock, "Note Concerning Human Respiratory Irritations Associated with High Concentrations of Plankton and Mass Mortality of Marine Organisms," *Journal of Marine Research* 7 (1948): 56–62.

9. "The Chemical Weapons Convention Declaration and Report Handbook, Jan. 2004"; Available at: www.cwc.gov.

10. See Dietrich Schlichter and Hans W. Fricke, "Coral Host Improves Photosynthesis of Endosymbiotic Algae," *Naturwissenschaften* 77 (1990): 447–450.

11. See Siro Kawaguti, "Effect of the Green Fluorescent Pigment on the Productivity of Reef Corals," *Micronesia* 5 (1969): 313; A. Salih, A. Larkum, G. Cox, M. Kuhl, and Ove Hoegh-Guldberg, "Fluorescent Pigments in Corals Are Photoprotective," *Nature* 408 (2000): 850–853; Sophie Dove, Ove Hoegh-Guldberg, and Shoba Ranganathan, "Major Colour Patterns of Reef-Building Corals Are Due to a Family of GFP-like Proteins," *Coral Reefs* 19 (2001): 197–204.

12. Ray Berkelmans, Glenn De'ath, Stuart Kininmonth, and William J. Skirving, "A Comparison of the 1998 and 2002 Coral Bleaching Events on the Great Barrier Reef: Spatial Correlation, Patterns, and Predictions," *Coral Reefs* 23 (2004): 74–83.

13. David R. Bellwood, Terence P. Hughes, Carl Folke, and Magnus Nyström, "Confronting the Coral Reef Crisis," *Nature* 429 (2004): 827–833.
14. Terence P. Hughes, Andrew H. Baird, David R. Bellwood, M. Card, Sean R. Connolly, Carl Folke, Rick Grosberg, Ove Hoegh-Guldberg, Jeremy B. C. Jackson, Joan Kleypas, Janice M. Lough, Paul Marshall, Magnus Nyström, Steve R. Palumbi, John M. Pandolfi, Brian Rosen, and Joan Roughgarden, "Climate Change, Human Impacts, and the Resilience of Coral Reefs, *Science* 301 (2003): 929–933.
15. D. Bryant, L. Burke, J. McManus, and M. Spalding, *Reefs at Risk: A Map-Based Indicator of Potential Threats to the World's Coral Reefs* (Washington, D.C.: World Resources Institute, 1998).
16. N. V. C. Poluninand C. M. Roberts, ed., *Reef Fisheries* (London: Chapman and Hall, 1996).
17. Roger Y. Tsien, "Rosy Dawn for Fluorescent Proteins," *Nature Biotechnology* 17 (1999): 957.

11. Lighting Up the Brain

1. Stanley Finger, *Minds behind the Brain: A Discovery of the Pioneers and Their Discoveries* (New York: Oxford University Press, 2000). p. 21.
2. Ibid., p. 36.
3. *A Collection of Several Philosophical Writings of Dr Henry Moore* (London, 1653), p. 34.
4. Nieves Fernandez and Caoimhghín S. Breathnach, "Luis Simmaro Lacabra [1851–1921]: From Golgi to Cajal through Simmaro, via Ranvier?" *Journal of the History of Neurosciences* 10 (2001): 19–26.
5. Marina Bentivoglio, "Life and Discoveries of Santiago Ramón y Cajal," Nobel website: http://nobelprize.org/medicine/articles/cajal.
6. Santiago Ramón y Cajal, *Recollections of My Life*, trans. E. Horne Craigie and Juan Cano (Cambridge: MIT Press, 1989), p. 308.
7. Ibid., p. 36.
8. Ibid., p. 144.
9. Ibid., p. 325.
10. Ibid., p. 325.
11. Camillo Golgi, "The Neuron Doctrine—Theory and Facts," in *Nobel Lec-*

tures, *Physiology or Medicine, 1901–1921* (Amsterdam: Elsevier Publishing Co., 1967), p. 192.

12. Julia Tsai, Jaime Grutzendler, Karen Duff, and Wen-Biao Gan, "Fibrillary Amyloid Deposition Leads to Local Synaptic Abnormalities and Breakage in Neuronal Branches," *Nature Neuroscience* 7 (2004): 1181–1183.

13. Linda Buck and Richard Axel, "A Novel Multigene Family May Encode Odorant Receptors: A Molecular Basis for Odor Recognition," *Cell* 65 (1991): 175–187.

12. Glowing Thoughts

1. Walter Hess, 1949 Nobel Lecture, "The Central Control of the Activity of Internal Organs," in *Nobel Lectures in Physiology or Medicine 1942–1962* (Singapore: World Scientific, 1999), p. 250.

2. José M. R. Delgado, *Physical Control of the Mind: Toward a Psychocivilized Society* (New York: Harper & Row, 1969), p. ix.

3. All information on the bull was obtained from the film *Radio Stimulation of Brave Bulls*, produced by José M. R. Delgado with the collaboration of Francisco J. Castejon and Francisco Santisteran.

4. John A. Osmundsen, "'Matador' with a Radio Stops Wired Bull," *New York Times*, May 17, 1965, p. A1.

5. Delgado, *Physical Control of the Mind*, p. 168.

6. José M. R. Delgado, "Permanent Implantation of Multilead Electrodes in the Brain," *Yale Journal of Biology and Medicine* 24 (1952): 351–358.

7. Delgado, *Physical Control of the Mind*, p. 137.

8. Peter R. Breggin, "The Return of Lobotomy and Psychosurgery," remarks presented by the Honorable Cornelius E. Gallagher of New Jersey in the House of Representatives, *Congressional Record*, Feb. 24, 1972, vol. 118, part 5, pp. 5567–5577.

9. Ibid., p. 5574.

10. Delgado, *Physical Control of the Mind*, pp. 222–223.

11. See the DARPA website, http://www.darpa.mil.

12. Sanjiv K. Talawar, Shaohua Xu, Emerson S. Hawley, Shennan A. Weiss, Karen A. Moxon, and John K. Chapin, "Behavioural Neuroscience: Rat Navigation Guided by Remote Control," *Nature* 417 (2002): 37–38.

13. Hannah Hoag, "Neuroengineering: Remote Control," *Nature* 423 (2003): 796.

14. Interview with Patrick Nagle, March 23, 2005. Unless otherwise indicated, further quotes from and information about Nagle come from this interview.

15. Spinal Cord Injury Information Network (www.spinalcord.uab.edu).

16. Interview with Mikhail Shapiro, March 3, 2005.

17. Micah S. Siegel and Ehud Y. Isacoff, "A Genetically Encoded Optical Probe of Membrane Voltage," *Neuron* 19 (1997): 735–741.

18. Kazuto Ataka and Vincent A. Pieribone, "A Genetically Targetable Fluorescent Probe of Channel Gating with Rapid Kinetics," *Biophysical Journal* 82 (2002): 509–516.

19. Gero Miesenböck, Dino A. De Angelis, and James E. Rothman, "Visualizing Secretion and Synaptic Transmission with pH-sensitive Green Fluorescent Proteins," *Nature* 394 (1998): 192–195.

20. Santiago Ramón y Cajal, *Histology of the Nervous System of Man and Vertebrates,* vol. 1, trans. Neely Swanson and Larry W. Swanson (Oxford: Oxford University Press, 1995), p. 3.

21. Osamu Shimomura, lecture entitled "Discovery of Aequorin and GFP," Aug. 29, 2004, Friday Harbor, Wash.

Acknowledgments

Like the scientific pursuit of fluorescent proteins, this book was a collaborative effort. We wish to thank the many scientists who contributed to the making of this book. At Harvard University Press, we are grateful to our editor, Nancy Clemente, for her sharp and seasoned editorial eye, and for always demanding more; and to Ann Downer-Hazell for approaching us to write this story and then carefully guiding us along the path to publication. We are indebted to the staff of the Institute for Marine and Coastal Sciences at Rutgers University (especially Judy and Fred Grassle and Gary Taghon) and the John B. Pierce Laboratory and Yale University for providing support and creative scientific environments. K. Buckley Gdula and Armelle Casau expressed early interest in this project and provided invaluable research. In addition, historical research for this book would not have been possible without the staff of the Rutgers University and Yale University libraries. Editorial expertise was provided by Lynnora Geoghegan, Steve Ives, Alfonso Serrano, and Audra Wolf. We thank Andrew Berlin, Tom Bibby, K. Buckley Gdula, Alex Kahl, L. Melnik, Susan Nakley, James Park, Alex Phipps, Maria Stoian, Steve Tuorto, Julia Tsai, and Bas van de Schootbrugge for insightful comments on numerous drafts of this book. Dan Tchernov, a friend and colleague of both authors, provided stimulating discussions on many aspects of coral biology. A warm thanks to the faculty and staff of Columbia University's Graduate School of Journalism, and to Sig Gissler's "Section Nine" for ringing reminders of the importance of eating oatmeal and burning shoe leather.

The National Institutes of Health and the National Science Foundation sponsored much of the research reported in this book. We are obliged to the staff and volunteers of the Earthwatch Institute (especially Mary Blue Magruder and Lotus Vermeer) for making possible our quest for fluorescent proteins on Lizard Island, Australia. And a special thanks to Marlaine, Edward, Lori, and Wyatt; Julia, Frances, and David.

Index

Micronesia, 175
Microscopes, 87–88; epifluorescent, 90, 91; fluorescent, 90, 91, 124–125; high-grade light, 90; laser scanning confocal, 90–91; multiphoton fluorescent, 91; multiphoton, 92; confocal, 92–93; standard light, 120; electron, 206
Mind control, 223
Minsky, Marvin, 90
Modified Green Fluorescent Protein, 143
Molds, 11; *Dicytostelium*, 148
Molecular biology, 97, 99, 100, 113, 122, 127, 133, 158, 164, 165–166, 228; techniques, 103, 160
Molecular biotechnology, 168
Molecules, fluorescent, 88, 89
Moline, Marksalps, 26
Mollusks, 11
Mombaerts, Peter, 210
More, Henry, 202
Morgan, Thomas Hunt, 30–32
Morin, James, 106, 149
Muller, Hermann, 31
Muscles, 157; calcium levels in cells, 71; cells, 155; movement controlled by the brain, 225–226
Mutation, 31, 142–143, 158–159; as form of resistance, 113; of proteins, **143**, 144, 149, 150, **156**, 184–185
Myosin light chain kinase, 155

Nagasaki, Japan, 49, 50, 51
Nagle, Matthew, 223–225
National security issues, 221
Nature (Biotechnology), 150
Nature (journal), 98, 142, 150, 175
Nematocysts, 182
Nematodes. *See* Worms
Neon paints, 87

Nervous system, 205, 210, 224
Neurological conditions, 223
Neurons, 125, **227;** in brain cells, 202, **205,** 206, **209,** 211, 215; properties of, 216–217; damage to, 226
Neurophysiology, 228
Neuroscience, 205, 206, 208
Neurotoxins, 191
Neurotransmitters, 216
Neutrons. *See* Subatomic particles (electrons, neutrons, protons)
New Caledonia, 175
New Guinea, 36
Newton, Sir Isaac, 78, 81, 82
Nightsea, Inc., 176
Ninhydrin (Luminol), 87

Observer's paradox, 83
Olfactory system, 208–212, **211, 230**
On Color (Aristotle), 11
"On Some Cases of Fluid Motion" (Stokes), 78–79
"On the Change of the Refrangibility of Light" (Stokes), 81
Opticks (Newton), 82
Optics, 91, 227
Organelles, **22**
Oxygen, 54; required for bioluminescence, 13, 69; molecular, 69

Pancreas, 157
Paralysis, 223, 224–225
Patents, 143, 153, 165; disputes over, 121–122
PCR technique, 172
Peyssonnel, Jean André, 188–189
pH, 68
Philips, George, 150
Phillips, C. E. S., 175